恭祝万达集团
成立 25 周年

2010
万达商业规划
WANDA COMMERCIAL PLANNING

中国建筑工业出版社

广场金街景观小品

室内步行街中庭采光顶

《万达商业规划2010》编委会
THE EDITORIAL BOARD OF WANDA COMMERCIAL PLANNING 2010

主编单位
Chief Editor
万达商业规划研究院
Wanda Commercial Planning & Research Institute

规划总指导
General Advisor in Planning
王健林
Wang Jianlin

编委
Executive Editors
赖建燕 黄大卫 朱其玮 叶宇峰 王元 冯腾飞
Lai Jianyan, Huang Dawei, Zhu Qiwei, Ye Yufeng, Wang Yuan, Feng Tengfei

参编人员
Editors
马红 刘冰 孙培宇 莫力生 李峥 范珑 刘阳 万志斌 熊伟 王群华 侯卫华 王鑫
毛晓虎 郝宁克 袁志浩 阎红伟 黄引达 耿大治 黄勇 张振宇 王雪松 谢冕
高振江 孙辉
Ma Hong, Liu Bing, Sun Peiyu, Mo Lisheng, Li Zheng, Fan Long, Liu Yang, Wan Zhibin, Xiong Wei, Wang Qunhua, Hou Weihua, Wang Xin, Mao Xiaohu, Hao Ningke, Yuan Zhihao, Yan Hongwei, Huang Yinda, Geng Dazhi, Huang Yong, Sun Zhengyu, Wang Xuesong, Xie Mian, Gao Zhenjiang, Sun Hui

李峻 梅咏 孙多斌 杨旭 田杰 朱莹洁 李斌 门瑞冰 常宇
Li Jun, Mei Yong, Sun Duobin, Yang Xu, Tian Jie, Zhu Yingjie, Li Bin, Men Ruibing, Chang Yu

校对
Proofreaders
兰峻文 张涛
Lan Lunwen, Zhang Tao

英文翻译及校对
Translators and Proofreaders
吴昊 张震 沈文忠 梅林奇 宋锦华 刘佩
Wu Hao, Zhang Zhen, Shen Wenzhong, Mei Linqi, Song Jinhua, Liu Pei

WANDA **2010**

创新是核心竞争力的前提
INNOVATION IS THE PREREQUISITE FOR THE KERNEL COMPETITIVE POWER

创新的根本是机制创新。企业创新要想形成持续，最根本的是要解决机制创新，靠机制推动而不是靠外力推动创新才能持续。

In principle, the innovation of enterprise organization is the root of the innovations. If any enterprise want to keep having a sustained innovation, above all, which should find the solution for the innovation of organization, the sustained drivers only promoted by internal organization, not external forces.

创新的核心是思想创新，就是解放思想，敢于求异思维。

The kernel of innovation is innovation of thought, which means to emancipate mind, with a bold differentiated thinking.

创新要合理适度。不能为了创新而创新，不能过度创新；不能觉得一切创新就是好事，要看适不适合自己的情况，适不适合市场的情况。

Making innovations need to be an appropriate and rational scale. We can not only search for innovations just for innovations, or overextended innovations as well. Meanwhile, not all innovations are good things, which really depends on whether it is suitable for the reality and market situation.

创新是一个持续完善的过程，需要在实践当中不断完善、不断琢磨。

Making innovations is a sustainable development process. Which need to keep improving and studying.

——王健林董事长，《创新是核心竞争力的前提》主题演讲
Wang Jianlin, theme speech:
Innovation Is the Prerequisite for the Kernel Competitive Power

商业设计不同于建筑设计，一定要请懂商业的人来搞设计。设计中要关注几个要点：

Commercial design is different from normal architecture design, the former must invite professional commercial designer to participate. Please notice some key points as below:

第一，交通优先。设计一定首先考虑车辆从哪进哪出，一定要考虑周边道路的关系，交通能迅速疏散，能进来。

Firstly, the traffic structural system is the top priority. The beginning planning should fully consider the vehicle entrance and exit, related circulation and relationship with nearby street, which should be a well planned evacuation flow.

第二，出入口的选择。做一个大型购物中心，不是一个门，应该是七八个门。要考虑最主要的出入口应该放在哪儿，哪里是人们特别愿意走的方向。

Secondly, making the right choice of entries. To set up a large shopping mall, not only a single one but at least 7 or 8 main necessary entries, which need to consider where the most important one should be set up, and which route is preferable to the customers.

第三，关于人流的动线。人的步行是有固定习惯的，人们在购物的时候，疲劳的尺度也是非常有规律的。一般来说，步行街的长度最好在400米之内。另外，路线的规划，我们万达商业规划研究院有一句名言叫"不走回头路！"。

Thirdly, planning the right pedestrian circulation. The people have fixed habits for walking, when shopping in the mall, the customers also have the regular fatigue degree. Normally, it is better to keep the length of pedestrian street under 400m. Regarding the pedestrian circulation planning, Wanda Commercial Planning & Research Institute have a famous phrase saying: "we never return to the same road we come !".

万达将会编写一本关于购物中心的投资和建设方面的书籍，我们愿意把我们的经验总结出来，公开发表。关于设计有专门一个详细的章节去讲应该怎么设计。

Wanda is going to issue a book on the investment and construction of shopping mall, and we are delighted to share and publish our experience. Regarding the design, it will be explained in detail in a specified chapter.

——王健林董事长，万达集团2012商业年会高端论坛
Wang Jianlin
Chairman of Board and President of Wanda Group
at 2012 Annual Meeting

万达集团董事长
王健林
Wang Jianlin
Chairman of Board and President of Wanda Group

万达集团董事长王健林指导规划院工作
Wang Jianlin guiding Wanda Commercial Planning & Research Institute

目 录
CONTENTS

万达商业规划2010	012	WANDA COMMERCIAL PLANNING 2010
万达广场方案设计	014	SCHEMATIC DESIGN FOR WANDA PLAZA
万达酒店设计简述	018	WANDA HOTELS AND RESORTS DESIGN BRIEF
万达广场	023	PART 1 WANDA PLAZAS
合肥包河万达广场	024	HEFEI BAOHE WANDA PLAZA
广州白云万达广场	036	GUANGZHOU BAIYUN WANDA PLAZA
福州金融街万达广场	050	FUZHOU FINANCIAL STREET WANDA PLAZA
宜昌万达广场	064	YICHANG WANDA PLAZA
包头青山万达广场	072	BAOTOU QINGSHAN WANDA PLAZA
济南魏家庄万达广场	080	JINAN WEIJIAZHUANG WANDA PLAZA
武汉菱角湖万达广场	084	WUHAN LINGJIAOHU WANDA PLAZA
呼和浩特万达广场	090	HOHHOT WANDA PLAZA
绍兴柯桥万达广场	098	SHAOXING KEQIAO WANDA PLAZA
无锡滨湖万达广场	104	WUXI BINHU WANDA PLAZA
天津河东万达广场	112	TIANJIN HEDONG WANDA PLAZA
襄阳万达广场	116	XIANGYANG WANDA PLAZA
宁波江北万达广场	120	NINGBO JIANGBEI WANDA PLAZA
长春红旗街万达广场	124	CHANGCHUN HONGQIJIE WANDA PLAZA
沈阳铁西万达广场	130	SHENYANG TIEXI WANDA PLAZA
酒店	139	PART 2 HOTELS
合肥万达威斯汀酒店	140	THE WESTIN HEFEI BAOHE
襄阳万达皇冠假日酒店	146	CROWNE PLAZA XIANGYANG
福州万达威斯汀酒店	152	THE WESTIN FUZHOU MINJIANG
无锡万达喜来登酒店	158	SHERATON WUXI BINHU HOTEL
宜昌万达皇冠假日酒店	162	CROWNE PLAZA YICHANG
导向标识精选	169	PART 3 SELECTION OF GUIDING SIGNS
项目索引	179	PART 4 INDEX OF THE PROJECTS

万达商业规划2010
WANDA COMMERCIAL PLANNING 2010

In 2010, the Institute formulated and published 44 operational standards and technical standards. And for the first time, it made 17 Group standards, mainly on the basis on Wanda Plazas, including *Wanda Construction Standards 2010*, *Guidelines for Energy Saving at Wanda Shopping Centers*, *Standards and Guidelines for Performance-Based Fire Safety Design*, *Wanda's Standards for Controlling the Designing of Curtain Walls* and *Standards for the Design of Direction Signs in Underground Parking Area* and soon. These standards enable the planning and designing of Wanda Plazas to proceed on the tracks of modern corporate systems, which laid an integral solid foundation for the holistic quality upgrading of Wanda Plazas in 2011.

2010年,万达商业规划院组织编纂的《万达广场规划设计管控要点》正式颁布执行。该《管控要点》对万达商业综合体在规划设计过程中的各项设计要求进行了整理,共分11个专业321个要点。《万达广场规划设计管控要点》对于指导设计单位高效完成符合集团要求的设计任务提供了前提条件,同时对于项目的建设实施也有很好的指导、检查作用。

In 2010, *Key Points on the Controlling of the Planning & Design of Wanda Plazas*, compiled by the Institute, was formally published and executed. It streamlined various designing requirements in the course of planning and designing of Wanda's commercial complex, with 321 key points of 11 specialties. This provided a precondition for directing the designers to efficiently fulfill designing tasks in accordance with the Group's requirements. At the same time, it can be used to direct and check the construction of projects.

2010年,万达商业规划研究院与公安部天津消防研究所合作,开展了"大型综合性商业建筑防火设计关键技术研究"的科研工作,共同编制《大型综合性商业建筑设计防火规范》。该规范的编制完成对国内大型综合性商业建筑的防火设计起到极为重要的指导和规范意义,填补了目前国内大型综合性商业建筑防火设计的空白。

In 2010, the Institute cooperated with Tianjin Fire Research Institute of the Ministry of Public Security in the research on "Key Technical Studies on the Fire Safety Design of Large Comprehensive Commercial Buildings" and jointly compiled the *Fire Prevention Norms on Large Comprehensive Commercial Buildings*. The compilation of the Norms played an extremely important directional and regulatory role to the fire prevention design of domestic large comprehensive commercial buildings.

2010年,万达商业规划研究院牵头完成的《万达集团"绿色、低碳"战略研究报告》,首次公开明确地提出了万达集团商业建筑"绿色、低碳"战略目标,即:2011年及以后开业的项目均达到绿色建筑一星设计标准;2011年至2015年间已开业项目逐年降低运行能耗2%~3%;2013年取得5个项目绿色建筑一星运行标识认证;2015年实现运营管理水平均达到绿色建筑一星运行标准。

In 2010, *Wanda Group's Report on the "Green & Low Carbon" Strategic Research* clearly proposed the "Green & Low Carbon" strategic objective of Wanda Group's commercial building publicly for the first time. The strategic objectives are: all projects opened for business in and after 2011 shall reach the designing standard

万达集团董事长王健林指导规划院工作
Wang Jianlin guiding Wanda Commercial Planning & Research Institute

2010年是万达集团商业地产大幅提速发展的一年。2010年开业的万达广场共15座,比2009年开业的8座翻了一番,是2008年开业的2座万达广场的8倍,酒店的开业数量更是从2009年的2个激增到8个。

The year of 2010 witnessed a substantial growth in Wanda Group's commercial real estate business. 15 Wanda Plazas were opened this year, nearly doubling the number in 2009, namely, 8. It is also 8 times the number opened in 2008, that is, 2. The number of hotels opened for business is also increased to 8, from the number in 2009, namely, 2.

2010年是万达商业规划研究院的标准化年。万达商业规划院自2007年,经过2008年、2009年两年的摸索与积累,完善了自身的组织架构并开始逐渐成熟走向创新。万达商业规划研究院成为集团当年颁布标准最多的单位。

2010 is also the year of standardization for Wanda Commercial Planning & Research Institute. With its trials and accumulation in 2008 and 2009, the Institute refined its own organizational structure and began to display its maturity and innovation. The Institute became the department that published most standards within the Wanda Group that year.

2010年,万达商业规划研究院共组织制定和颁布实操性技术标准44项,主持完成了《万达2010版建造标准》、《万达购物中心节能工作指南》、《消防性能化标准及工作指南》、《万达幕墙设计管控标准》和《地下停车场导向标识设计标准》等集团标准17项。使万达广场的规划设计在现代企业制度化轨道上运行发展,为2011年万达广场全面的品质提升奠定了不可或缺的坚实基础。

327 个计划管控模块节点
327 Scheduled models

of 1-star green building; projects already opened for business between 2011 and 2015 shall lower their operational energy consumption by 2%~3% year by year; 5 projects shall be awarded "1-star green building operation" certification label in 2013; the level of operation and management shall reach 1-star operational standard for green buildings in 2015.

2010年是万达绿建节能划时代的一年。该年开业的万达广场中,有3个万达广场首次获得了国家住房和城乡建设部绿色建筑设计标识,实现了中国商业类建筑设计与建造"绿建节能"零的突破。其中,广州白云万达广场获得二星级"绿色建筑设计标识",武汉菱角湖万达广场、福州金融街万达广场获得一星级"绿色建筑设计标识"。此三个项目在2012年又获得了"绿色建筑运行一星认证标识",再次实现了中国商业建筑"绿色运营"零的突破。从2010年到2012年,仅两年时间,万达商业建筑的低碳节能,便实现了从设计建造到管理运营的真正意义上的全寿命周期的绿色建筑!

The year of 2010 is an epoch-making year for Wanda's green building and energy conservation. Among all Wanda Plazas opened that year, 3 was awarded Green Building Star Certification Label by the Ministry of Housing and Urban-Rural Development, which is the very first breakthrough for China's green commercial building. Among the awarded projects, Guangzhou Baiyun Wanda Plaza was rated as "2-star Green Building Designing Label" and Wuhan Lingjiaohu Wanda Plaza and Fuzhou Financial Street Wanda Plaza were rated as 1-star buildings of "Green Building Designing Label". These three projects also won the "1-star Green Building Operation" certification label, which means another first breakthrough in the "green operation" of commercial buildings. Within only two years, from 2010 to 2012, the low-carbon energy conservation of Wanda's commercial buildings realized the whole life cycle of green buildings in the real sense, from designing and construction to management and operation!

2010年也是万达集团的标准化年。2010年,大连万达商业地产股份有限公司组织完成了"项目管理计划模块",将商业地产的多部门、多链条、多交叉的复杂全产业管理全程纳入计算机信息化管控,从项目的前期挂牌、公司组建、总图规划、成本测算,到项目初期的方案设计、报批报建、施工图设计、招商组织、财务运作,再到项目中期的土建装修、机电安装、招商签位、消防论证、营销操作、中期检查,至项目后期的整改调试、环境美化、绿建申报、各类验收、竣工复盘,项目管控全程共涉及集团几大系统、十多个部门,总计327个计划节点。"项目管理计划模块"信息化系统的应用,使万达集团同时管理当年开业及后两年开业的多达几十个在建设项目成为可能,也使万达商业规划院每年设计并进行设计管控上百个项目成为可能!万达商业规划院的标准化是万达集团的标准化体系的一部分。万达"项目管理计划模块"信息化管理,使万达的商业地产管理水平提升到全球领先水平!

2010 is also the year of standardization of Wanda Group. In 2010, Dalian Wanda Commercial Estate Co., Ltd. produced the "Schedule Module for Project Management", to incorporate the complex whole-industry management involving multiple departments, multiple chains of management and multiple overlapped businesses into computerized and informationized control, that is to say, from the listing, establishment of the company, master planning and cost estimation in the preliminary phase, to the schematic designing, submission of reports for approval and construction application, designing of the construction drawings, organization of investment promotion and financial operation in the initial phase, from civil construction and decoration, installation of mechanical and electrical equipment, completion of the layout map of businesses, fire prevention argumentation, marketing operation and checking in the middle phase to the rectification & commissioning, environmental decoration, green building application, various acceptance tests, readjustment after completion and project controlling in the late period of project. Altogether several major systems of the Group and over 10 departments are involved, with 327 scheduled milestones in total. The application of the "Schedule Module for Project Management" information system enables Wanda Group to manage as many as dozens of projects under construction or designing at the same time, which are to be opened in the current year and future two years. And it is even possible for Wanda Commercial Planning & Research Institute to design and control the designing of over a hundred projects each year! The standardization of the Institute is also one part of thee standardization system of Wanda Group. The informationized management of Wanda's "Schedule Module for Project Management" upgraded Wanda's commercial real estate management to the top level in the world!

万达集团执行董事、总裁丁本锡(中)、大连万达商业地产股份有限公司执行总裁齐界(左一)等集团领导对规划院工作进行现场指导
The executive director / president of Wanda Group Ding Benxi (M), CEO Qi Jie (L1) guiding Wanda Commercial Planning & Research Institute

广州白云万达广场
获得二星级绿色建筑设计标识
Guangzhou Baiyun Wanda Plaza awarded "2-star" green building design certification label

广州白云万达广场
获得一星级绿色建筑运行标识
Guangzhou Baiyun Wanda Plaza awarded "1-star" green building operation certification label

2010年也是万达集团重视品质,万达广场开始提升产品品质的初年。万达集团在2000年步入商业地产10年后,首次对全年开业的万达广场进行品质评比。2010年开业的部分万达广场,如合肥包河万达广场、广州白云万达广场,率先突破了多年雷同的效果形式,唤醒了项目的品质意识,为2011年及2012年的全面创新及全面品质提升,积累了从设计到管理的宝贵经验。可以说,2010年的标准化建设,是万达商业规划2011年品质提升及2012年创新突破的坚实基础。

The year of 2010 is also the first year that saw Wanda Group attach great importance to quality and upgrade the product quality of Wanda Plaza. After 10 years since Wanda Group entered commercial real estate in 2000, for the first time, a quality competition was launched for all Wanda Plazas opened in the whole year. Some Wanda Plazas opened for business that year, such as Hefei Baohe Wanda Plaza and Guangzhou Baiyun Wanda Plaza, took the lead in breaking through the usual form of effects used for numerous years in the past, which awakened the awareness of quality for projects. This accumulated valuable experience from designing to management for the all-round innovation and holistic quality enhancement in 2011 and 2012. It is safe to say that the standardization in 2010 laid a solid foundation for the quality enhancement in 2011 and innovative breakthrough in 2012 of Wanda's commercial planning.

大连万达商业地产股份有限公司高级总裁助理
万达商业规划研究院院长
赖建燕
Lai Jianyan
Senior Assistant of the President of Dalian Wanda Commercial Estate Co., LTD
President of Wanda Commercial Planning & Research Institute

万达广场方案设计
SCHEMATIC DESIGN FOR WANDA PLAZA

商业地产项目的方案设计至关重要，是商业地产项目投资和运营成败的关键！

Schematic design is of utmost importance to a commercial real estate project and is the key to the success of commercial real estate project investment and operation!

商业地产项目的方案设计最核心是动线设计！它决定了商业广场生或死，旺或衰，好或坏。商业有一句话，叫"隔街死"，还有一句话，叫"一步差三成"，讲的就是商业动线对商业地产项目的极端重要性。万达集团在万达广场的发展历程中也有过深刻经验教训——沈阳太原街万达广场：由于动线规划不合理，导致太原街万达广场人气很差、商铺经营惨淡，最后不得不花比原投资更高的代价回购并重新规划建设。

The crux of the schematic design for a commercial project lies in the circulation design, which constitutes the matter of life or death, prosperity or depression, good or bad. There are two aphorisms in the business world, one goes that "although one side of a street is bustling, the other side may be deserted"; and the other goes that "though only a few steps from each other, one shop is crowded while the other is empty." Both of the biting aphorisms tell us how important the circulation design is to commercial projects. In the development of Wanda Plazas, Wanda Group learnt a profound lesson in this respect: due to a flawed circulation design, our Taiyuan Street Wanda Plaza in Shenyang was quite unpopular, making it hard for the stores in it to have a good business performance. In the end, Wanda Group had no choice but to repurchase the area for a complete re-profiling and reconstruction, which incurred more cost than the original investment.

最早期购物广场的动线均为"哑铃"型。通过在"哑铃"的两头设置主力店，中间布置小商铺并自然形成商业街，使行走其中的购物人群自然流动起来，商气和商机随之滚滚而来。"哑铃"型动线也成为商业动线最早的雏形。后来，随着商业购物广场的发展，又产生了各式各样商业动线类型。

The circulation of the earliest shopping plazas is in "dumbbell shape". The two anchor stores at each end of the "dumbbell" link up all the in-between smaller stores into a commercial street naturally, introducing increased density of traffic flows into their vicinity and attracting waves of business opportunities. The "dumbbell shape" has become the earliest form of commercial circulation. Later on, various other forms of commercial circulation emerged along with the development of commercial shopping plazas.

图1："哑铃"型动线商业平面图
Figure 1 Plan of a Commercial Real Estate Project with a Dumbbell Shape Circulation

图2："哑铃"型变形动线商业平面图
Figure 2 Plan of a Commercial Real Estate Project with a Variant Dumbbell Shape Circulation

万达广场从"哑铃"型动线又发展出以"U"字型、"一"字型等为主的基本人流动线。如图2，它与"哑铃"型动线相比步行街与主力店互动性更强，均好性更好。万达广场动线的最大特点是：动线简单！动线简单让消费者很容易记住商铺的具体位置，不容易迷失方向；动线简单、不走回头路、没有死角和分叉，让消费者在逛街中自然而然必经每家商铺，使商家均好性和利益最大化。动线越简单商场越安全、商业越成功！经研究，国内成功的购物中心，均属于动线简单型，如北京新光天地、深圳万象城、上海IFC、广州太古汇等。

On the basis of the "dumbbell shape" circulation, Wanda Plaza developed other basic forms of pedestrian circulation mainly featuring U shape and straight line shape, as shown in Figure 2. The strengths of these forms, compared with the "dumbbell shape" circulation, are the intensified interactivity between pedestrian streets and anchor stores and a better sharing of resources. Simplicity is the hallmark of Wanda Plaza circulation, making it easy for customers to remember the exact location of stores and thus do not easily get lost; simple circulation avoids such things as turning back, dead corners or bifurcation, so that customers unwittingly stop by each store. Therefore, resources are better shared and business profits are maximized. The simpler the circulation is, the safer the plaza is and more successful the business is! Studies have shown that all the successful shopping centers in China are built on a simple circulation model. Examples include Shin Kong Place in Beijing, Wanxiang Shopping Mall in Shenzhen, IFC in Shanghai, Taiguhui in Guangzhou, and so on.

万达广场平面设计的特点可以用"一街带多楼"来概括。"一街"即一条室内商业步行街，"多楼"即百货楼、娱乐楼等。这是万达广场多年成功实践经验的总结和高度概括，也是万达广场一切平面设计的基础。

"One street promotes numerous buildings" could be used to sum up the features of the plane design of Wanda Plaza. "One street" means one indoor commercial pedestrian street and "Numerous buildings" include department store building recreation building, and so on. Actually, it is a conclusion and the highest generalization of successful experience in developing Wanda Plaza for many years. It is counted as the foundation for all the plane design for Wanda Plazas.

"一街"是一条三层楼高，约300米长，带玻璃采光顶的室内商业步行街。一层到三层分别布置时尚服饰、生活配套、餐饮酒楼，是万达广场的精华。

3 stories high and about 300m long, "one street" is an indoor pedestrian street with a glass daylighting roof. From the first to the third floor are fashion stores, life supporting facilities and restaurants, the gem of Wanda Plaza.

"多楼"包括百货楼、娱乐楼、超市楼等，是万达广场体量最大、人流量最多的主力店楼群；特别是娱乐楼，是万达广场人气最旺的场所，它和步行街顶层餐饮酒楼一同起到拉动万达广场人气的重要作用。

Consisted of a shopping department, a recreation building, a supermarket, etc., "numerous buildings" are regarded as the biggest anchor store buildings with the largest size and the heaviest pedestrian flow in Wanda Plaza. The recreation building is a particularly popular place, which plays a crucial role along with the roof restaurants in the pedestrian street in attracting customers to Wanda Plaza.

图3: 万达广场商业平面示意图
Figure 3 Diagrammatic Plan of Wanda Plaza

图4: 万达广场剖面示意图
Figure 4 A Cross Section Diagram of Wanda Plaza

same in business. The most crowded and the most popular place is always the first floor! But from the lower to the higher floors, the density of crowds decreases. Given this, the theory of "water towers" and "water wheels" should be applied. Pedestrian flows are taken to the top floor of the pedestrian street or the recreation building by escalators and elevators. Varied types of dining businesses and Wanda Cinema could attract endless streams of customers to higher floors. Then, pedestrian flows can be directed from the higher floors to the lower ones just like water running down.

"漩涡"原则。能量在娱乐楼和百货楼两大"漩涡"中心聚集，步行街连接其间，这样，人流流动起来；中庭作为"漩涡"中心，必然向周围发散能量；所以，中庭聚集步行街最主要商家、万达百货的主入口等。

There is another theory—the theory of "whirlpools". Energy gathers up in the center of the two "whirlpools" — the recreation building and the shopping department which are connected by the pedestrian street, allowing pedestrians to flow through unimpededly. As the center of a "whirlpool", the atrium surely gives off energy. Therefore, the atrium is surrounded by the most important businesses and the chief entrances to Wanda Dept. Store.

万达广场通过一条四季如春的室内商业步行街，将几大主力店楼串起来。同时，通过地上、地下几十部水平和垂直扶梯和电梯，将来自四面八方的人流迅速送达购物、餐饮和娱乐目的地。

The anchor store buildings in Wanda Plaza are linked up by an indoor commercial pedestrian street, which offers constantly comfortable temperature all year round. In addition, dozens of horizontal or vertical escalators and elevators on and under the ground swiftly take flows of people from all directions to their destinations for shopping, dining or recreation.

万达广场的动线设计功能分工明确，流线设计合理。分为客流动线、车流动线、货流动线、员工动线；几种动线交通互不干扰、互不交叉，组成城市综合体高效动线交通体系。

The circulation design of Wanda Plaza boasts clearly defined functions and perfectly designed flow lines. There are circulations of customers, vehicles, goods and staff, which do not interfere or overlap with one another and form an efficient urban complex traffic system.

客流动线
Customer Circulation

万达广场客流动线原则可用"水论商法"来形象描述：

The theory of the customer circulation of Wanda Plaza could be vividly captured by the phrase "business rules of the water":

"塔提"、"车扬"原则。像古代用水塔或水车将水提到高处，水就能自然从高处往低处流一样；商业也一样，首层人流最多、人气最旺！但越往上走，人流越少、人气越差；我们应用"塔提"、"车扬"原则，通过扶梯和电梯将人流拉到步行街和娱乐楼顶层，通过餐饮业态和万达影城等娱乐业态将人流源源不断吸引上来；然后，人流又像水流一样，从高处自然流向低处。

The theory of "water towers" or "water wheels" is adopted. In ancient times, water towers and water wheels were used to lift water so that it could flow down naturally. The case is just the

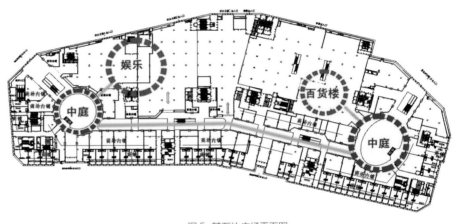

图5：某万达广场平面图
Figure 5 Plan of a Wanda Plaza

车流动线
Vehicle Circulation

机动车流线是万达广场动线设计的重点。它的主要设计特点是：小客车及货车严格分流、互不干扰；小车方便停放、就近原则；停车场方便识别，方便寻找；机动车快速离开原则。万达广场是国内大型购物广场中最早引进弱电智能化停车系统的企业。

As a focal point of the circulation design for Wanda Plaza, the vehicle circulation has the following main design features: mini-passenger cars and vans are strictly divided to avoid jamming; small cars can be parked easily in accordance with the proximity principle; parking lot is easy to tell and find; and vehicles can leave speedily. Wanda Plaza is one of the earliest large shopping plazas in China that introduced the intelligent weak-current parking system.

货流动线
Goods Circulation

货流动线有几种，一种是各主力店及室内步行街的进货动线，主要通过各主力店地下层的货运电梯实现；另一种是各主力店的垃圾动线，垃圾直接运至地下干、湿垃圾间，集中存放运输。这两种动线使用时段不同，互不干扰。

There are two types of goods circulation. One is to bring goods to anchor stores as well as the indoor pedestrian street. Goods are mainly transported through the underground freight elevators of anchor stores. The other is to take out garbage from anchor stores. The garbage are directly transported to and stocked in the underground garbage rooms for solid and fluid waste. The two types of circulation run independently during different periods of time.

员工动线
Staff Circulation

万达广场内员工包括商管人员及各业态的营业人员。商管人员因为工作时间与万达广场营业时间不同，需要单独设置员工上下班通道。一般通过靠近城市干道的娱乐楼或者百货楼的电梯通往办公区域。三层货运通道是餐饮工作人员及垃圾的独立通道。

The staffs of Wanda Plaza include management staff and sales staff from various types of businesses. Since the management staffs' work time does not follow the open time of Wanda Plaza, an independent staff passage should be set up. Usually, elevators of the recreation building or the shopping department that is close to thoroughfares lead to office areas. The three story freight passage is reserved for catering staffs as well as garbage transportation.

"一个万达广场，一个城市中心"，"万达广场让生活更美好"。万达广场是集购物、餐饮、娱乐、生活配套等为一体，既满足市民购物消费、餐饮娱乐的需要，更成为市民聚会、休闲的场所。万达广场已经成为很多城市儿童、青年、成年及老年人喜爱的公共活动场所，成为城市"公共客厅"，名副其实"24小时不夜城"！

"Wanda Plaza is the heart of the city", "Wanda Plaza makes our life better". Wanda Plaza is a place that integrates shopping, dining, entertainment, recreation and life supporting, not only meeting the needs of citizens for shopping, dining and recreation but also functioning as a venue for citizens to gather and relax. Wanda Plaza now has become a public space and "living room of the city" popular with kids, youngsters, adults and elders in many cities and is worthy of the name —"the city that never sleeps".

万达集团从2000年开始率先进入中国商业地产领域，为中国商业地产的高速发展和探索做出了自己的贡献和努力。万达集团今年将跨入世界级企业行列，成为世界自持物业第二的国际企业，我们将继续为提高城市综合体和购物广场的规划设计水平而努力。

In 2000, Wanda Group pioneered into China's commercial real estate sector, contributing to the rapid development and exploration of China's commercial real estate. This year Wanda Group will rank among world-class enterprises and as the No.2 international enterprise with the most self-sustaining properties. We will make a constant effort to improve the planning and design of urban complexes and shopping plazas.

万达商业规划研究院副院长
叶宇峰
2012.12.8

Ye Yufeng
Vice President of
Wanda Commercial Planning & Research Institute
December 8th, 2012

万达酒店设计简述
WANDA HOTELS AND RESORTS DESIGN BRIEF

万达集团目前已开业28家五星和超五星级酒店。万达集团计划到2015年开业70家五星和超五星级酒店，营业面积300万平方米，成为全球最大的五星级酒店业主。万达拥有五星级酒店品牌——万达嘉华、超五星级酒店品牌——万达文华以及顶级奢华酒店品牌——万达瑞华。

Wanda Group has opened 28 five-star and super five-star hotels. By 2015, the Group plans to run 70 five-star and super five-star hotels covering an area of 3,000,000 m^2, making itself the largest owner of five-star hotels in the world. Wanda Group owns its 5-star hotel brand — Wanda Realm, super 5-star hotel brand — Wanda Vista and top luxury hotel brand — Wanda Reign.

万达集团旗下酒店分为城市类酒店和度假类酒店两大类。万达城市类酒店通常位于城市中心，一般和万达广场等业态组成万达城市综合体。万达度假类酒店通常位于文化旅游区，一般和旅游小镇、秀场、主题公园等组成万达文化旅游项目。

There are two types of hotels run by the Wanda Group: urban hotels & resorts. Urban hotels are usually located in the city center and, together with various types of businesses in the Wanda Plaza, they are a part of the urban complex. Resorts, on the other hand, are often located in a cultural tourism zone and are a part of Wanda's cultural tourism program, together with tourist towns, show areas and theme parks.

万达酒店
Wanda Hotel

城市酒店服务的是各类人群，位于城市中心的位置，建筑风格一般都与城市相融合。城市酒店作为综合体各业态中的一个重要组成部分，提供了吸引、蓄存并提供客源的功能。酒店结盟城市综合体，能使酒店借助地产项目完善的商业商务配套提升其商业价值，地产则可借助酒店优良的产品抽象提升硬件和服务的质量与档次。在布局上，酒店需要在与综合体其他功能部分有机联系的同时，也保持自己相对的独立性和私密性。因此，酒店在综合体中所处位置、与综合体其他组成部分的关系，以及由酒店通往综合体的连接方式，也因综合体项目主题和定位的变化而各有千秋。酒店与综合体的关系是相辅相成的。在对酒店或综合体进行设计管理时，使酒店成为综合体有机的一部分，独立而又统一，最大限度地发挥综合体的特点和优势，以达到双赢的目的。

Providing services to people from all walks of life, the urban hotels in the central part of a city are often designed in a style that fit into the context of the city. As an integrated part of the urban complex, these hotels have served to attract and accommodate people, as well as provide sources of customers for other businesses. A combination of the hotel and the complex has a two-way effect: with the good business facilities provided by the real estate project, the commercial value of the hotel will be increased; at the same time, with the hotel's excellent customer services images, the real estate will be able to improve the quality of its hardware and services. In terms of the overall arrangement, the hotel needs to be connected with other functional parts of the complex, and at the same time maintain its independence and privacy. Therefore, a hotel will be wholly different from another in its location, its connection with the other parts of the complex and the way it links to the complex, depending on the different theme and positioning of the complex project. The hotel and the complex are inseparably interconnected. Therefore in the design and management of the hotel or the complex, the hotel should be regarded as an organic part of the complex and is supposed to be in harmony with the complex while keeping its unique features, in order to maximize the advantages of the complex to achieve a win-win outcome.

度假酒店是以接待休闲度假游客为主，为休闲度假游客提供住宿、餐饮、娱乐与游乐等多种服务功能的酒店。度假酒店大多建立在有旅游资源的区域，提供度假资源，例如海滨、山地、林地、湖泊、温泉等自然景区，并与自然风光相融合，注重客房景观。

The resorts mainly serve people who are on holiday and having a leisure time, providing them a range of services, such as accommodation, catering, entertainment and amusement. Most of them have been built in a place with tourism resources, such as the seaside, mountains, woodlands, lakes and hot springs. Merging into the natural scenery, they offer rooms with a splendid view of the outside.

1 万达城市类酒店主要技术要求
1 The technical requirements for Wanda's urban hotels

1.1 酒店大堂
1.1 The hotel lobby

大堂是酒店中最重要的区域，是酒店整体形象的体现，同时酒店大堂也是一个酒店对外展示自我的平台。万达酒店大堂的显著特点为两层通高，瑞华酒店首层层高为9.0米，挑空区净高大于12米；文华酒店及以下为7.0米，挑空区净高大于10米。酒店大堂面积在700~1000平方米之间，形成万达酒店空间宏伟，气势庞大的第一印象。

As the most important component of a hotel, the hotel lobby reflects the overall nature of the hotel; meanwhile, it also serves as a platform for the hotel to show itself. A notable feature of Wanda's hotel lobby is that two storeys of the building have been carved out to create a grand entrance to the hotel. The first floor of Wanda Reign is 9.0m high, and the carved area is more than 12m high. For the Wanda Vista and the hotels in a lower rank, the first floor is 7.0m high, and the carved area is more than 10 m high. Wanda's hotel lobby is 700 ~1000m^2, creating a magnificent first impression.

1.2 酒店餐厅
1.2 The hotel restaurant

餐饮经营收入弹性大，在酒店整体收入中占有很大比重，因此，餐饮空间设计在酒店总体设计中有很重的分量。酒店中的餐饮空间，一般包括中西餐厅、特色餐厅、酒吧、咖啡厅等。设计上考虑空间座位容量及形式，餐桌混合比例，餐桌及服务通道规格等。根据中国文化特点和市场特色，万达酒店强调餐饮包间明显多于散座空间，常在5:1以上。散座空间所有的中餐大包房若空间容许，均设置卫生间和步入式衣帽间。万达所有项目中的餐厅包房都设有沙发区，有送餐间、卫生间。特色餐可与中餐厅同在一层，也可独立设置。餐厅厨房面积一般不小于餐厅面积的30%。

Income from catering, which may fluctuate quite widely, accounts for a large proportion of the overall revenue of the hotel. So the layout of the dining rooms is of great concern to the whole design of the hotel. Generally, there is an array of dining rooms in a hotel, including Chinese style dining room, Western style dining room, special cuisine dining room, bar and cafe. There are a lot of things to consider in designing the dining area: the capacity and types of seating, the ratio of the chairs to tables, table sizes, the size of the servants' corridor, and so on. In light of the characteristics of the Chinese culture and the market, the private rooms are far more than those provided for individual customers in Wanda's hotels (often with a ratio of 5 to 1). Every private room that serves Chinese food in the individual customer zone, if space is enough, has a washroom and a walk-in closet. The private rooms in all kinds of Wanda hotels each have an exclusive area for sofa, a food delivery room and a washroom. Special cuisine dining room and Chinese style ding room can either be arranged on the same floor, or be arranged separately. The area of a kitchen is normally no less than 30% of that of the dining room.

1.3 酒店会议室
1.3 The hotel meeting room

会议室数量及面积，应根据酒店面积规模相应设置。会议层可与宴会厅设在同一楼层，不与其他功能合设在同层以避免相互干扰。万达酒店会议室多数是100平方米左右，小会议室50~60平方米。会议室均设有衣帽间与卫生间。

The number and size of the meeting room should be designed according to the size of the hotel. The meeting room and the banquet hall, excluding the rooms with other functions, may be located on the same floor, in order to avoid interference. Most of the meeting rooms in Wanda's hotels are about 100 m^2, while the small meeting rooms are about 50 to 60 m^2. All of the meeting rooms have a closet and a washroom.

1.4 酒店康体中心
1.4 The hotel recreation center

康体中心含健身房、游泳池、各类球场、棋牌室、舞厅、KTV等内容。其中酒店的类型、等级、经营项目的不同，使之配备面积比也不同。度假酒店中设置比例相对会高一点。不符合酒店定位和需求的建议不宜硬性设置，避免造成不必要的成本浪费。康体娱乐项目区域适宜设置在酒店低层的裙楼或地下室等位置。与客房区域要保持一定距离，避免对客房造成声音和人流的干扰和影响。

The recreation center includes gym, swimming pool, the playfield for all kinds of ball games, chess room, ballroom and KTV. Hotels of different types, ranks and operating items may have differently-sized recreation centers. For the resorts, the recreation center is relatively larger. To avoid unnecessary cost, we should not impose rigid requirements without looking at the positioning and needs of the hotel. It is proper to locate the recreation center on the lower floors of the hotel or the basement. And it should not be too close to the guest rooms to avoid noise and interference effects.

1.5 酒店客房
1.5 The hotel guest room

万达五星标准客房间面积40~42平方米左右，客房层层高3.8米。六星酒店客房或酒店标准间客房面积大于45平方米时设有步入式衣帽间。酒店套间均设步入式衣帽间，套间客房的卫生间设有两个洗手盆。

A standard guest room in Wanda's five-star hotel is about 40 to 42 m^2 and at 3.8 m high. A guest room or standard room in Wanda's six-star hotel is larger than 45 m^2 and has a walk-in closet in it. All suites in Wanda's hotel have a walk-in closet and there are two wash basins in the bathroom of the suite.

1.6 行政层及总统套
1.6 The administration floor and the presidential suite

酒店行政层设置两个部长套间，一室两厅，且至少有一个套间应与一侧的标准间通过客厅双向双门连通。行政酒廊设置6~7个自然间，行政酒廊层和总统套房层高5.1米。总统套房根据地域级别的不同，面积在250~400平方米。总统套房的主人房和夫人房分别对公共走廊设置单独入口，设计为男女主人房可单独销售的形式。与总统套房相连的套间，装修风格、色彩、材质等标准与总统套房一致。

On the administration floor of the hotel there are two minister suites, which have one bedroom and two living rooms. They are designed in such a way that at least one of them is adjoined to the neighboring standard room through a two-way sliding door in the living room. Along the administration corridor, there are 6~7 normal rooms. The administration floor and the presidential suite are of 5.1 m high. The presidential suite is about 250 to 400 m^2, depending on the different situations in various regions. It is designed in such a way that both the host-room and the wife-room have a separate entrance on the public corridor, so that the two rooms can be used independently by a man or a woman. The decorating style, color and materials of its neighboring suites are of the same standard as those of the president suite.

2 万达度假类酒店主要技术要求
2 The technical requirements for Wanda's resorts

万达度假酒店群的特点是集中设置，根据自身所在地的资源环境基础，提炼出最有用的优势资源，并利用自然景观进行场地造景，从而形成度假酒店的最直接形象。所有酒店建筑围绕景观伸展布局，实现充分利用景观资源。室内外形成组团，领域性清晰。度假酒店大堂与室外之间实现景观上的通透，其特色餐厅数量比城市酒店多，会议室数量相对较少。酒店客房面积通常比城市酒店客房面积大，且多设置景观露台或阳台，增加客房观景面。客房尽量不用隔墙，部分室内功能空间融合或简化，增强空间尺度感。景观浴室和卫生间设计，提升度假生活的舒适感。

The hotel buildings of Wanda's resorts are concentrated. They are designed on the basis of the resources of their surrounding environment. To make the best use of these resources, a natural resort landscaping is created to make the resort stand out. All the built architectures are adjoining to the landscape, so that they can make full use of the landscape resources. The resort gives the indoor and outdoor equal accommodation for scenery. Sitting in the lobby, one can have a splendid view of the landscape outside. Compared to the urban hotels, the resorts have more special cuisine rooms and fewer meeting rooms. In the resorts, guest rooms are usually larger than those in the urban hotels, and most of them have a terrace or balcony for the enjoyment of the views. Inside the guest room, partitions are reduced to the largest extent and some functional parts are also combined or simplified to make the room look more spacious. Landscape elements are added to the bathroom and toilet for people to have a more comfortable experience in the resort.

万达度假村
Wanda Holiday Village

当前，中国酒店业仍处于一个品牌发展的上升时期，尽管国内有近百万家不同类型与级别的酒店企业，但是值得推崇的品牌还是太少，而万达酒店致力于成为"值得推崇"的品牌之一。万达酒店以其独特的方式与中国特色将与万达集团一起走向世界巅峰。

At present, the hotel industry of China is still in the period of brand development. There are nearly one million hotels of different types and ranks in China, but few are award-wining brands. Wanda Hotels and Resorts, however, is going to become one of the "award-wining" brands. In its own unique way and with the Chinese characteristics, Wanda Hotels and Resorts, together with the Wanda Group, will strive to ascend to the top of the world.

万达商业规划研究院副院长
刘冰
Liu Bing
Vice President of Wanda Commercial Planning & Research Institute

PART 1 万达广场
WANDA PLAZAS

合肥包河
万达广场
HEFEI BAOHE WANDA PLAZA

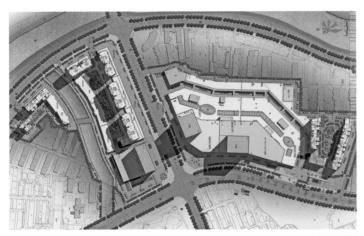

总平面图

合肥包河万达广场建筑面积约70万平方米，占地面积11.1公顷，位于安徽省合肥市包河区美菱旧厂区地块，东至巢湖路，南至现状地界，西至马鞍山路，北至现状地界。作为合肥"十一五"重点项目，在政府的大力支持下，万达集团响应城市大发展的时代主题，致力于推动城市商业的不断演变，完成在合肥首个力作——包河万达广场。

Hefei Baohe Wanda Plaza is located in Meiling old factory plot, Baohe District, Hefei, east to Chaohu Rd, south to the status boundary, west to Maanshan Rd, north to the status boundary, with the gross floor area of about 700,000 m^2, and the overall site area of 11.1 ha. As the 11th Five-Year Plan key project of Hefei, in response to the times theme of urban development under strong government support, Wanda Group devoted itself to promote the urban commercial continuous evolution, complete the first masterpiece in Hefei Baohe Wanda Plaza.

合肥包河万达广场全景

主外立面

合肥包河万达广场是全业态的第三代商业综合体，由购物中心、威斯汀酒店、超高层甲级写字楼、室外步行街、滨河酒吧街、超高层豪宅、普通住宅等组成；购物中心引进GUCCI、万宝龙、登喜路等国际一线品牌，这是安徽首个最具规模和品牌实力的大型城市综合体，对于整合城市资源、提高城市商业格局，具有划时代的影响力。

Hefei Baohe Wanda Plaza is the third generation urban complex, consisting of shopping mall, Westin hotel, high-rise grade A office building, outdoor pedestrian street, the riverfront bar street, the high-rise mansion and ordinary residential etc.; The shopping mall imported GUCCI, Montblanc, Dunhill and other world class brands. The large urban complex which possesses the largest scale and brand strength in Anhui province had the epoch-making influence for the integration of urban resources and raising the city's business Pattern.

大商业外观

大商业主入口

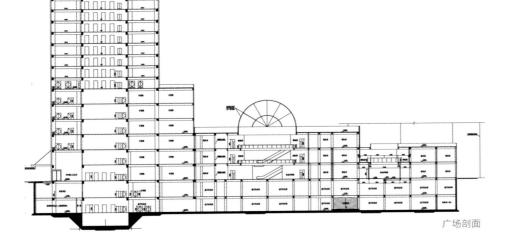

广场剖面

广场日景

室内步行街扶梯

1F

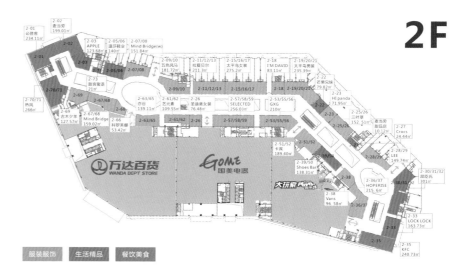

2F

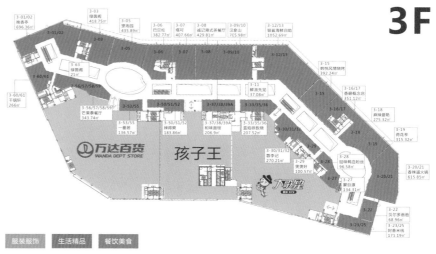

3F

品牌落位图

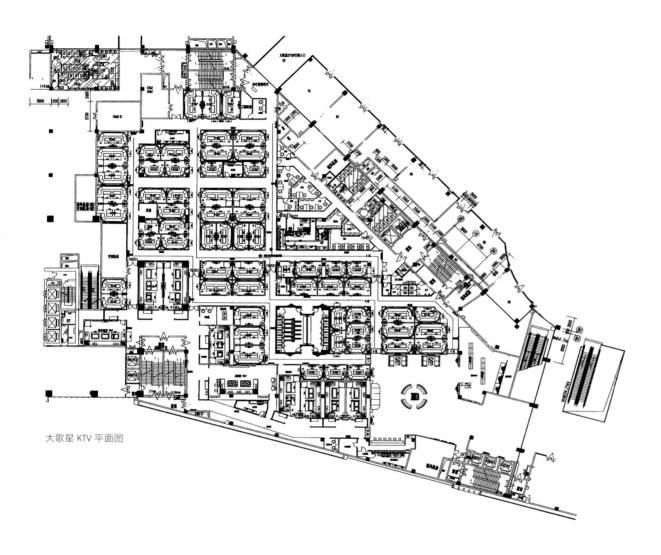

大歌星 KTV 平面图

室内步行街内装

室内步行街中庭采光顶

大商业外街夜景

室内步行街直街

商业入口广场

广州白云万达广场
GUANGZHOU BAIYUN WANDA PLAZA

广州白云万达广场是一座集商业中心、五星级酒店、商务酒店、室外步行街、甲级写字楼等业态为一体的城市综合体；位于广州白云区，东至云城东路，西至云城西路，南至横五路，北至白云路；总用地面积约21.1公顷，总建筑面积56.3万平方米。地上建筑面积44.4万平方米，其中购物中心9.6万平方米，写字楼7.7万平方米，五星级酒店3.7万平方米，室外步行街2.0万平方米，地下建筑面积11.9万平方米。

Guangzhou Baiyuan Wanda Plaza is an urban complex integrating commercial center, 5-star hotel, commercial hotel, outdoor pedestrian street, Class-A office building and so on. It is located in Baiyuan District, Guangzhou. It reaches east to Yuncheng East Road, west to Yuncheng West Road, south to Hengwu Road and Baiyun Road in the north. It occupies an area of 21.1 ha, with a GFA of 563,000 m². The floor area above the ground reaches 444,000 m². Among this, the shopping center occupies 96,000 m², the office building occupies 77,000 m², the 5-star hotel occupies 37,000 m², the outdoor pedestrian street occupies 20,000 m² and the underground floor area occupies 119,000 m².

广场全景

项目分为A、B、C三个区。其中A区含三栋甲级写字楼和一栋五星级酒店；B区为万达购物中心，业态包含万达百货、大玩家超乐场、大歌星KTV、万达影城、国美电器、儿童城、超市及餐饮名店和室内步行街招租业态。B区内设置两块11米×80米的大型高清天幕，炫彩烘托出购物氛围；C区为SOHO公寓。三个区通过城市步行连廊有机连接成整体。

The project is composed of Areas A, B and C. Area A has 3 Class-A office buildings and a 5-star hotel; Area B is Wanda shopping center. Its businesses include Wanda Dept. Store, Super Player Park, Super Star KTV, Wanda Cinemas, GOME, Children Town, supermarket, famous food & beverage brands and indoor pedestrian street for rental. Area B has two 11m × 80m HD overhead projected screens, whose colors offer an ideal atmosphere for shopping; Area C is the SOHO Apartment. The three areas are organically connected with urban pedestrian corridors.

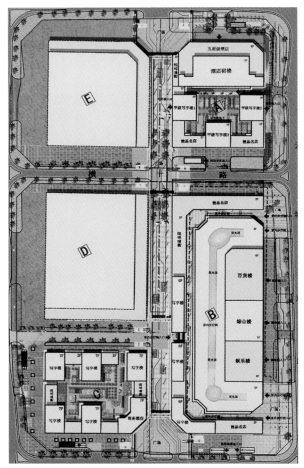

总平面图

广场喷泉夜景

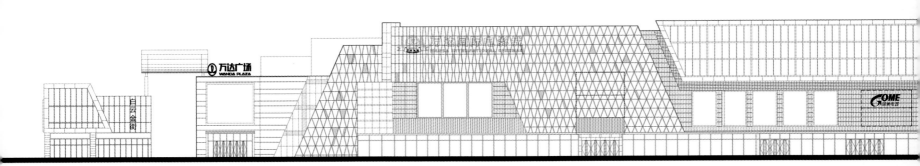

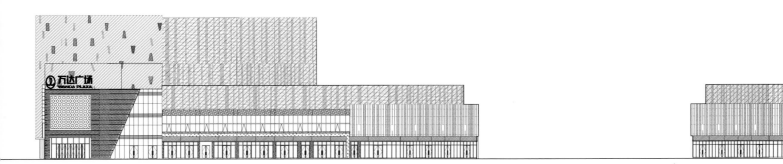

南北立面图

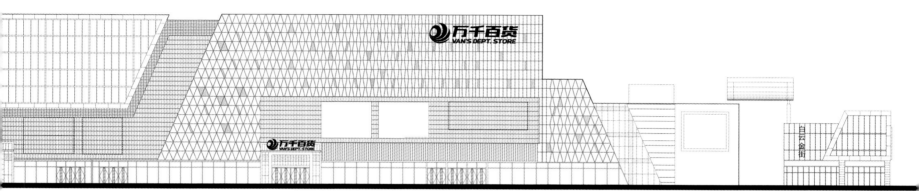

主立面图

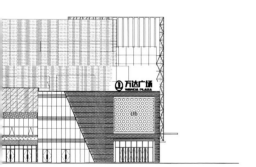

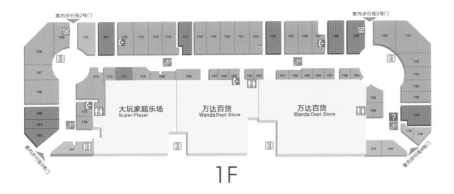

1F

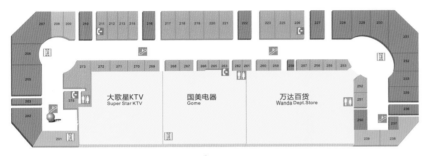

2F

3F

品牌落位图

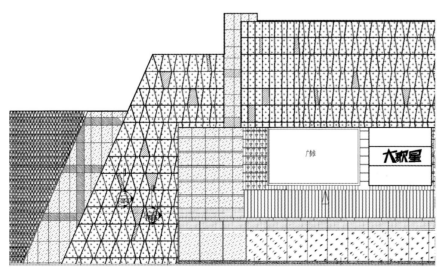

局部立面图

万达百货入口

外立面　　　　　　　　　　　　　　　　　　　　商业广场　　　　　　　　　　　　　　　　　　　　室外步行街

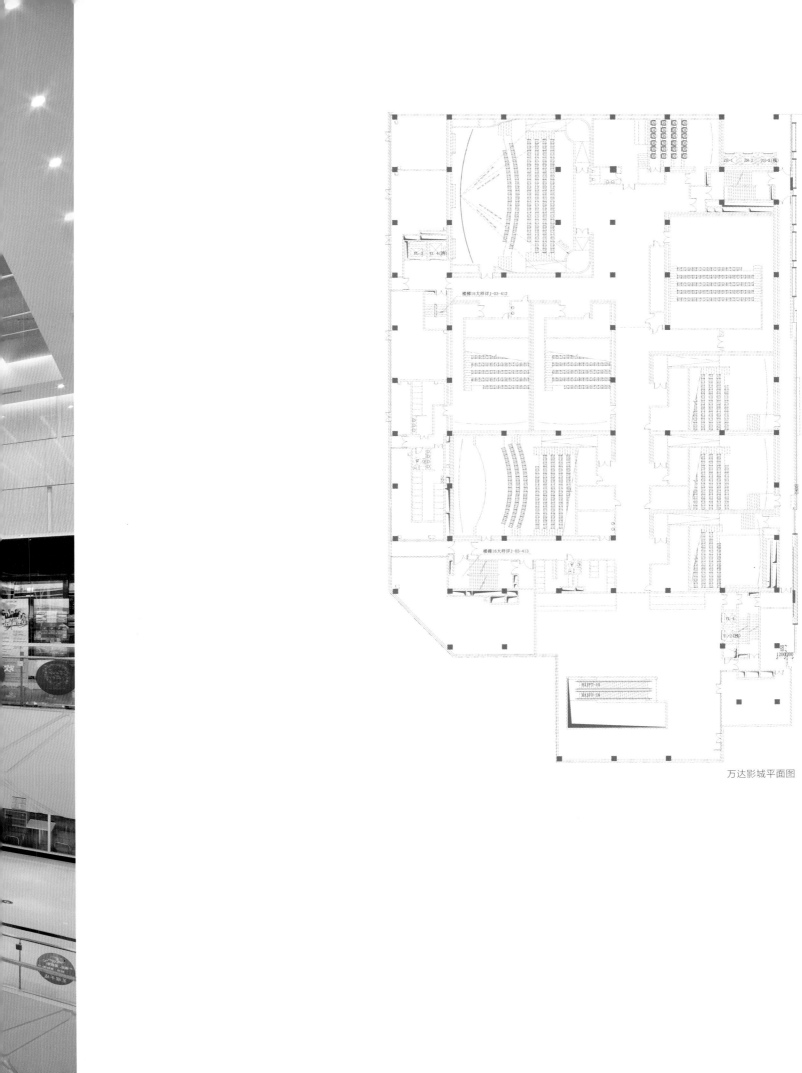

室内步行街扶梯

万达影城平面图

室内步行街入口

室内步行街休息区

室内步行街中庭

广场夜景

福州金融街
万达广场
FUZHOU FINANCIAL STREET WANDA PLAZA

福州金融街万达广场位于正在建设的海峡金融商务区核心区域，南临江滨中大道，北临鳌峰路，东为规划前横路，西为规划曙光路，未来地铁4号线穿过项目，交通十分便利。福州金融街万达广场开业后，商业中心、银行、保险等各种金融行业将相继进入，该区域已经逐步形成新的金融中心。

Fuzhou Financial Street Wanda Plaza is situated in core area of Strait Financial and Business District under construction and is bordered by Jiangbin Avenue to south, Aofeng Road to north, planning Qianheng Road to east and planning Shuguang Road to west, the Plaza arounds convenient transportation conditions as the planning Metro Line 4 will run through the Project in the future. With the business of commercial centre, Banks, insurance and other financial industries after the opening of Wanda Plaza, the region will become a new financial center.

广场鸟瞰效果图

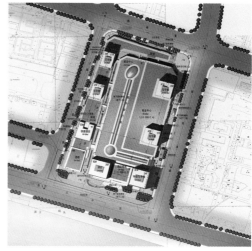

总平面图

福州金融街万达广场总占地约7.2公顷，总建筑面积约40万平方米，是一个集五星级酒店、甲级5A写字楼、大型高档购物中心、大型娱乐中心、时尚步行街、SOHO办公为一体的城市综合体。城市的所有精华功能集于一体，产生极大的吸金效应，形成新的城市中心，提升整个城市形象。福州金融街万达广场的崛起是大福州城市建设的一个新的里程碑，万达广场强大的城市引擎力，将带动这座古老而年轻的海峡都市向新的高度腾飞。

Wanda Plaza covers an area of about 7.2 ha with a gross floor area of approximately 400,000m², It's an urban complex integrating 5-star hotel, Class 5A office building, large and high-end shopping center, large recreational center, pedestrian street and SOHO office buiding, drawing all advantages of the city into one to engender tremendous attraction effects and form a new city center to enhance urban image. The rise of Fuzhou Financial Street Wanda Plaza is a new milestone in urban construction and will serve as a powerful engine for this ancient and young strait metropolis to take off to the next level.

广场东北角日景

广场夜景

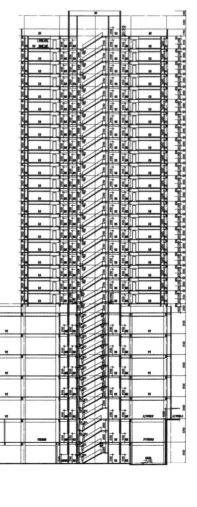

剖面图

广场夜景

室内步行街入口

广场东北角夜景

室外步行街夜景

外观细部

1F

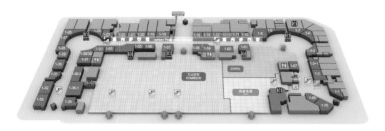

2F

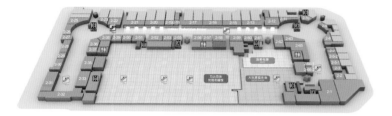

3F

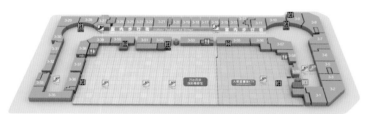

品牌落位图

室内步行街内景

室内步行街二层

室内步行街内景

室内步行街椭圆中庭

室内步行街

宜昌
万达广场
YICHANG WANDA PLAZA

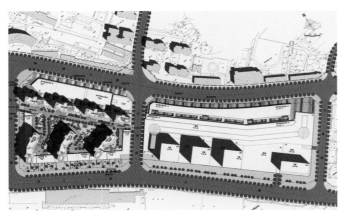

总平面图

宜昌万达广场坐落于宜昌市伍家岗区，位于沿江大道与夷陵大道之间，东临万寿路，南依沿江大道，西靠胜利一路，北至夷陵大道，堪称伍家岗区和西陵区的门户。

Yichang Wanda Plaza is located in Wujiagang District of Yichang. It is situated between Yanjiang Avenue and Yiling Avenue. The project borders east to Wanshou Road, south to Yanjiang Avenue, west to Shengli First Road and north to Yiling Avenue. It can be termed as the portal connection of Wujiagang District and Xiling District.

商业广场景观

室内步行街休息椅

室内步行街扶梯

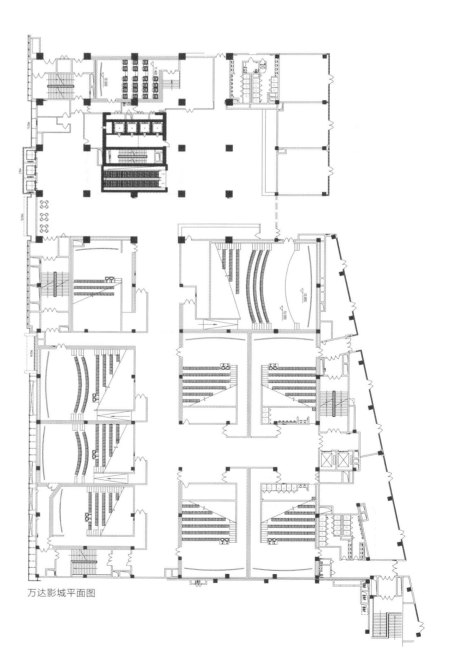

万达影城平面图

宜昌万达广场项目总占地8.5万平方米，总建筑面积47万平方米，是由多种不同形态的建筑产品构成的城市综合体，宜昌万达广场集万达百货、沃尔玛超市、室内商业步行街、室外商业步行街、五星级酒店、高档写字楼、精品住宅、万达影城、大型餐饮及休闲、娱乐健身为一体。项目周边建成及在建的大型居住区众多，路网发达，交通便捷，生活配套设施完备，拥有广阔的城市发展前景与潜力。

Yichang Wanda Plaza occupies an area of 84,900 m², with a GFA of 470,000 m². It is an urban complex composed of different types of architectural products. The project integrates Wanda Dept. Store, Wal-Mart Super Market, indoor commercial pedestrian street, outdoor commercial pedestrian street, 5-star hotel, high-class office building, boutique residences, Wanda Cinemas, large food & beverage outlets as well as leisure, entertainment and fitness facilities. There are numerous large residential areas either built up or under constuction around the project. The road network is developed, the traffic is convenient and supporting facilities for daily life is complete. All these can prove its broad prospects and potentials in the urban development.

整个项目以雄伟的姿态与创新的设计内涵集中体现了都市化、现代化的设计理念，它的建成将使所在商圈的商业档次及购物环境得到极大提升。

The whole project reflects its urbanized and modernized design concepts with its magnificent look as well as its design connotation featured by innovation. Its existence will substantially enhance the commercial quality and shopping environment of the business zone it is located in.

室内步行街中庭

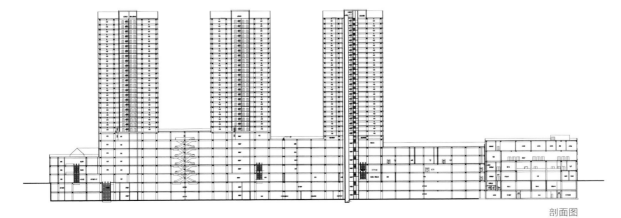

剖面图

室内步行街内景

室内步行街内景

1F

2F

3F

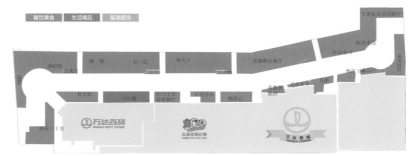

品牌落位图

室内步行街内景

外观全景

包头青山
万达广场
BAOTOU QINGSHAN WANDA PLAZA

广场日景

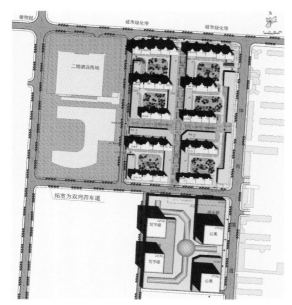

总平面图

包头青山万达广场位于包头市青山区，南临银河广场，东至体育馆路，西临香格里拉酒店，北至青年路。总占地面积10.6公顷，总建筑面积65万平方米，地上建筑面积54万平方米，其中住宅面积比例为45%，公建面积比例为55%。

Baotou Qingshan Wanda Plaza is located in Qingshan District, Baotou. Bordered by Yinhe Square on its south side, Stadium Road on its east side, Shangri-La Hotel on its west side and Youth Road on its north side, the project covers a total area of 106,000 m², and a gross floor area of 650,000 m², of which 540,000 m² is above ground. 45% of the project is residential and the rest 55% is taken up by commercial buildings.

该项目包括具有万达特色的大型商业综合体、高档写字楼以及高档住宅等多种业态，是集购物、娱乐、休闲、文化、居住等多种功能为一体的大型城市综合体项目。

The project consists of a large-size commercial complex with distinctive Wanda characteristic, high-end office building and high-end residential development, incorporating shopping, entertainment, leisure, culture, dwelling into one urban complex property.

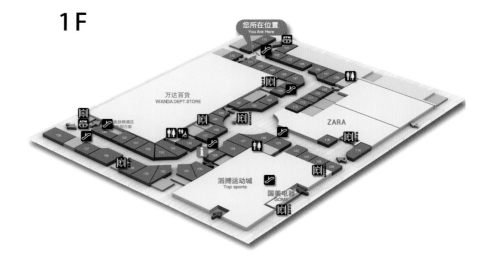

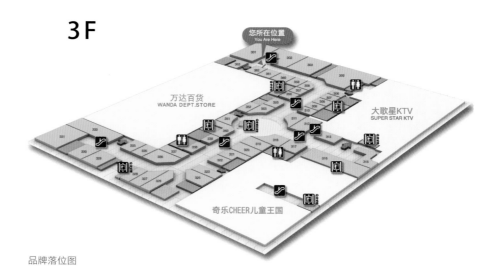

品牌落位图

室内步行街内景

室内步行街

万达影城大厅

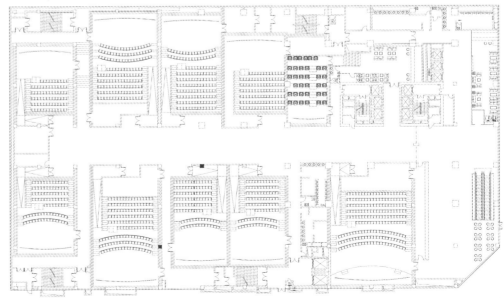

万达影城平面图

广场夜景

济南魏家庄
万达广场
JINAN WEIJIAZHUANG WANDA PLAZA

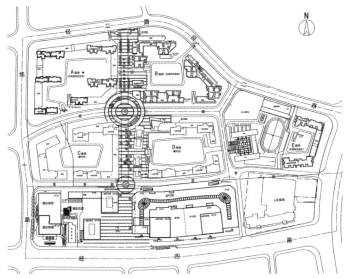

总平面图

本项目位于济南市市中区经四路以北，顺河街以西，经二路以南，纬一路以东。济南魏家庄万达广场总规划用地面积为23.0公顷，总规划建筑面积约为100.4万平方米。是由城市高尚社区、甲级写字楼、超五星级酒店和大型购物中心组成的城市综合体项目。其中沿经四路商业规划逾16万平方米，酒店5.1万平方米，写字楼15万平方米，精装公寓3万平方米。购物中心包含主力店有万达百货、国际连锁超市、国美电器、大玩家电玩城、大歌星KTV、万达影城。

The project is located to the north of Jingsi Road of Shizhong District of Ji'nan, the west of Shunhe Street, the south of Jing'er Road and the east of Weiyi Road. The master planning of Ji'nan Weijiazhuang Wanda Plaza has an area of 23.0 ha, while the master plan of its floor area occupies around 1,004,000 m². It is an urban complex composed of urban high-class community, class-A office building, super 5-star hotel and large shopping center. Among these, the commercial planning along Jingsi Road exceeds 160,000 m². The hotel area is 51,000 m², the office building area occupies 150,000 m² and the well decorated apartment area has 30,000 m². The shopping center has flagship stores in it, including Wanda Dept. Store, international chain supermarket, GOME, Super Player Park, Super Star KTV and Wanda Cinemas.

广场夜景

万达百货

室内步行街

室内步行街扶梯

武汉菱角湖万达广场
WUHAN LINGJIAOHU WANDA PLAZA

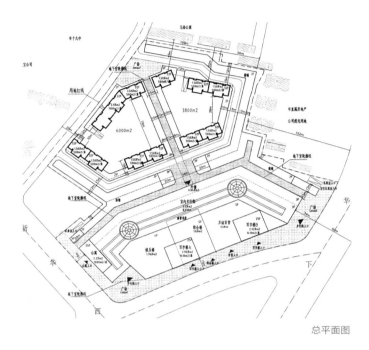

总平面图

武汉菱角湖万达广场位于武汉市江汉区，东临新华下路，西邻新华西路，南侧为规划道路、武汉新闻出版局，北侧为马场公寓、菱湖上品项目，西北侧为武汉市十九中。

Wuhan Lingjiaohu Wanda Plaza is located in Jianghan District, Wuhan. The plaza has its major frontage: east to Xin Hua Xia Rd, west to Xin Hua Xi Rd. south to Planning Rd & Publication Bureau, north side of Racecourse Apartments & Linghu Shangpin project, northwest side of Wuhan No.19th High School.

项目的总规划用地面积9.5万平方米，总建筑面积达51万平方米，由商业综合体、室外商业街、公寓、住宅及底商组成。其中大商业于2010年12月正式开业。项目地属武汉市中心地段，交通便利，具有浓郁的商业氛围，购物场所、文化教育、休闲等城市公用设施齐全，有良好的升值潜力。

The total planned area is 94,700 m², a total construction area of 510,000 m², by a commercial complex, outdoor commercial street, apartment, residence and business. In this area, major businesses officially opened in December 2010. The project is located in the center of Wuhan, with convenient transportation and strong business atmosphere. It has shopping mall, culture education, leisure space and other city public facilities, which has a good appreciation potentialty in the future.

广场全景

广场夜景

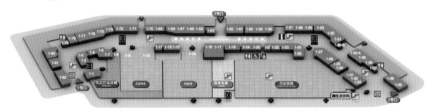
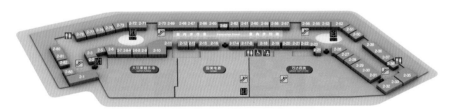
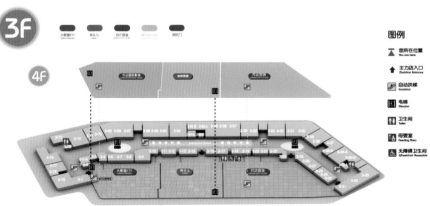

品牌落位图

室内步行街

呼和浩特
万达广场
HOHHOT WANDA PLAZA

广场日景

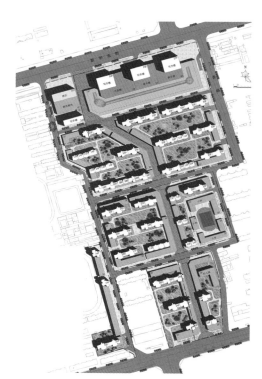

总平面图

呼和浩特万达广场是万达集团斥巨资为青城打造的开篇之作，占地面积约26.0公顷，总建筑面积将近130万平方米；占据城市主干道新华东街，扼守政治、经济、商业、文化四大中心交汇处；涵盖大型购物中心、五星级酒店、高档住宅、5A写字楼、商务公寓、商业街区等多重主流业态，集吃、喝、玩、乐、购一站式繁华配套于一身；并着力打造了当时万达最长的城市商业街，长达800米，全面满足市民日常生活所需，尽展世界潮流，旨在引领青城走向国际时尚最前沿。

Hohhot Wanda Plaza is an opening splendid project for Qing City with tremendous capital of Wanda Group, covers the site area of about 26.0 ha, and the gross floor area of nearly 1.3 million m^2. It occupies the city main roads-Xinhua Street which is the political, economic, commercial and cultural center of the four interchange. It covers multiple mainstream formats such as large shopping mall, 5-star hotel, high-end residential, Class 5A office building, business department, commercial block etc., integrating F&B, play, entertainment and shopping into one plaza and striving to build the longest urban commercial street with 800 m long to meet the public needs of people daily life comprehensively, spread the world trend and lead the city towards the fashion forefront.

室内步行街

广场景观小品

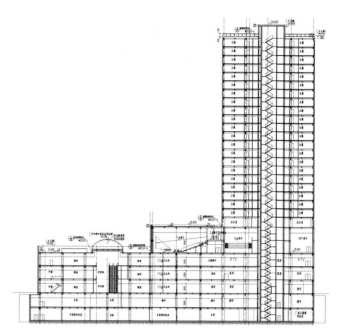

剖面图（一）

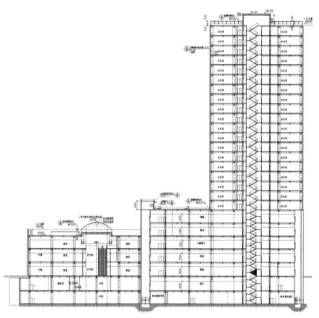

剖面图（二）

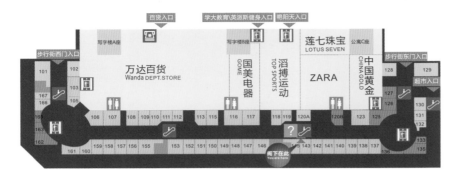

 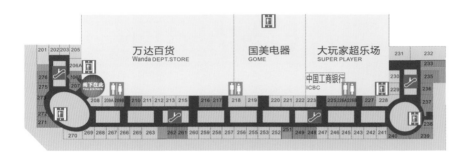

品牌落位图

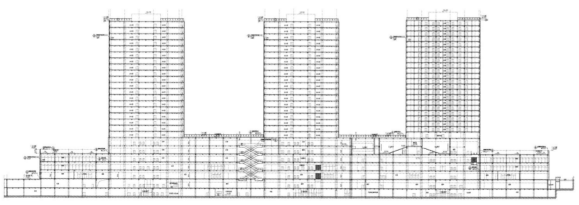

剖面图（三）

室内步行街扶梯　　室内步行街内景

室内步行街二层

绍兴柯桥万达广场
SHAOXING KEQIAO WANDA PLAZA

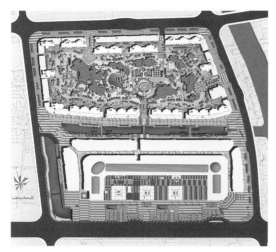

总平面图

绍兴柯桥万达广场位于绍兴县新城偏北，紧邻钱陶路商务办公带。项目北至华齐路，南至钱陶公路，西至湖中路，东至湖东路。绍兴柯桥万达广场的总规划用地面积18.2万平方米，规划总建筑面积约68.7万平方米。

本项目于2010年12月17日盛大开业，商业部分由购物中心、商务酒店、写字楼、室外步行街、沿河外铺组成。其中购物中心建筑面积17.4万平方米，地上10.4万平方米，地下7万平方米，商务酒店0.8万平方米，写字楼8.6万平方米，室外步行街2.6万平方米（全为地上面积），沿河外铺1.8万平方米。

Shaoxing Keqiao Wanda Plaza is located in Shaoxing County north to The New City, adjacent to the Qiantao Road commercial office zone. This project: north to the Huaqi Road, south to the Qiantao Road, west to Huzhong Road, East to the Hudong Road. Shaoxing Keqiao Wanda Plaza's total planning area is 181,800 m², a total planning construction area is about 687,000 m².

The project grandly opened on December 17th, 2010, whose commercial parts consist of shopping centers, trading hotels, office, outdoor pedestrian streets, which along the river outside the shop. The shopping center construction area is 174,000 m²: with the ground 104,000 m², underground 70,000 m²,commer cial hotel 8,000 m², office building 86,000 m²and outdoor pedestrian street 26,000 m²(all on the floor area). The river outside the shop is 18,000 m².

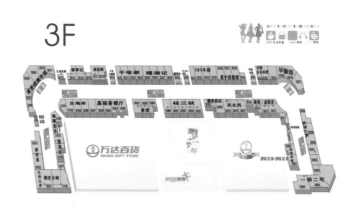

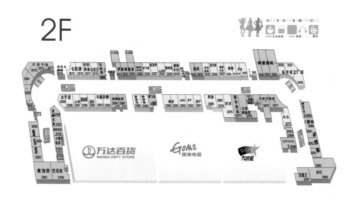

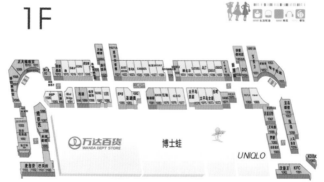

品牌落位图

广场室外夜景

广场外立面日景

广场室外夜景

广场景观

无锡滨湖
万达广场
WUXI BINHU WANDA PLAZA

广场西入口夜景

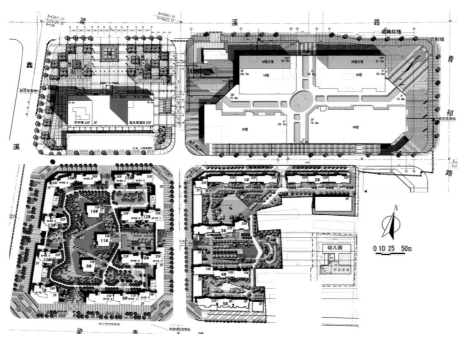

总平面图

无锡滨湖万达广场位于鱼米之乡江苏省无锡市，美丽的太湖之滨。

Wuxi Binhu Wanda Plaza is located in Wuxi, Jiangsu Province which is a fertile "land of fish and rice" along the beautiful Tai Lake.

项目占地约18.0公顷，总建筑面积约70万平方米，由万达商业综合体、喜达屋旗下喜来登5星级酒店、5A写字楼、高档住宅组成，其中：商业面积达12万平方米，分为四栋主力店楼，包含电器、健身、大型餐饮酒楼、万达百货、KTV、大玩家超乐场、儿童娱乐、大型超市等众多业态，构成丰富，种类齐全。

The site area of project is about 18.0 ha, and the gross floor area is near about 700,000 m^2. It is composed of Wanda commercial complex, 5-star Sheraton Hotel under Starwood Hotel Group, Class 5A office building and high-end residential. It covers commercial area of 120,000 m^2 which is divided into four buildings for the anchor stores, including multiple commercial patterns such as appliances, health & fitness, large restaurants, Wanda Dept. Store, KTV, Super Player Park, children's entertainment, and supermarket etc., with abundant composition and full range.

无锡滨湖万达广场2010年8月27日盛大开业，创造出江南又一璀璨明珠。

Wuxi Binhu Wanda Plaza was grandly opened on August 27th, 2010, which created another shining pearl within regions south of the Yangtze River.

广场日景

室内步行街中庭

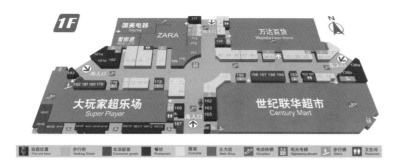

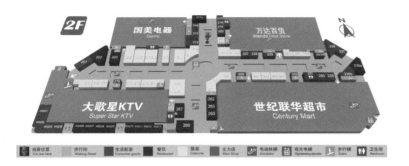

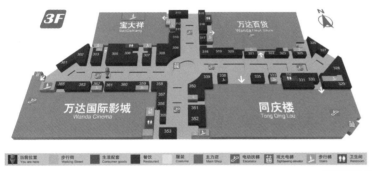

品牌落位图

室内步行街

室内步行街

室内步行街扶梯

室内步行街入口

天津河东万达广场
TIANJIN HEDONG WANDA PLAZA

津滨大道方向全景图

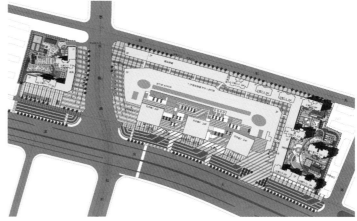

总平面图

天津河东万达广场位于天津市河东区津滨大道北侧，是市区去往机场的必经之地，扼守天津市区通往滨海新区之咽喉。整个地块主要划分为两个功能区，一是商业综合体，二是居住型公寓区。商业综合体位于用地中部，主要包括写字楼、百货楼、娱乐楼、综合楼、室内商业步行街、室外步行街（河东金街），建筑规模20.6万平方米。居住型公寓区，分为东、西两个地块，主要包括居住型公寓、底商式商业设施和少量公寓配套公建，建筑规模为20.4万平方米。

Tianjin Hedong Wanda Plaza is located in the Tianjin Hedong District, Tianjin Jinbin Avenue North, must be passed by from city central to airport, key spot between Tianjin downtown and the Tianjin Binhai New Area. The entire land is divided into two main functional areas, one is a commercial complex, the other is residential apartment area. Commercial complex is located in central, including office buildings, department store building, entertainment building, indoor pedestrian street, outdoor pedestrian street (East King Street), the total is 206,000 m². Residential apartment area, divided into two plots, including residence-apartment, commercial facilities and a small amount of facilities served for the apartments, the total residential floor area 204,000 m².

1F

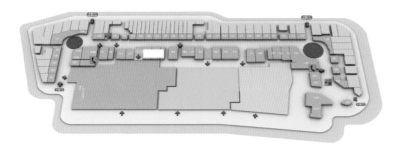

图例

2F

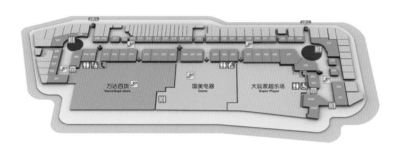

图例

3F

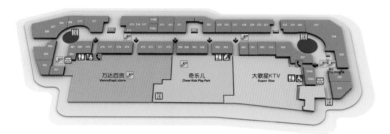

图例

品牌落位图

室内步行街中庭

襄阳
万达广场
XIANGYANG WANDA PLAZA

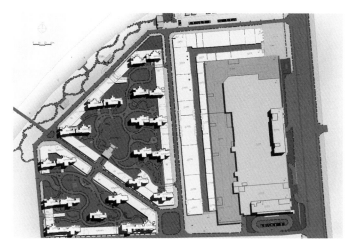

总平面图

襄阳万达广场于2009年8月开工，于2010年6月全面封顶，于2010年11月26日全面开业交付使用。主力店主要包含：万达百货、万达影城、国美电器、大歌星KTV、大玩家超乐场、华润超市、博士蛙、奇乐儿儿童主题公园杰之行、艳阳天酒家十大主力店，总建筑面积60万平方米，是襄阳首个城市综合体项目。

Xiangyang Wanda Plaza started in Augest, 2009, and its main work was finished in June, 2010. It was finished on Nov. 26th, 2010, which includes the 5-star hotel, and shopping mall. The main shop mainly contains: Wanda Cinemas, GOME, Super Star KTV, Super Player Park, Huarun supermarket, Dr. Rana, Cheers Kid's theme park Jie Zhixing, Yan Yangtian Restraunt ten main stores, with a total construction area 600,000 m^2, which is the first city complex project of Xiangyang.

购物中心室外日景

襄阳万达广场总占地13.3万平方米，其中大型购物中心约15万平方米，五星级酒店约4万平方米，甲级写字楼约3.5万平方米，室外步行街约3.1万平方米、城市中心住宅约25万平方米，项目车位总数2250个。广场通过各主力店与中小店铺的有机相联，引导商业中心顾客合理流动，满足消费者休闲、购物、娱乐为一体的"一站式消费"需求，成为商业中心的灵魂与纽带。

Xiangyang Wanda Plaza covers a total area of 133,000 m², of which a large shopping center of 150,000 m², 5-Stars Hotel of about 40,000 m², class A office building about 35,000 m², the outdoor pedestrian street of about 31,000 m², city center housing approximately 250,000 m². the project has 2,250 cars parking areas. The major stores are organiclly connected with the small shops, which guides customers effectively around the commercial center. The project meets consumers' leisure, shopping, entertainment demands and will become an "one-stop shopping commercial center".

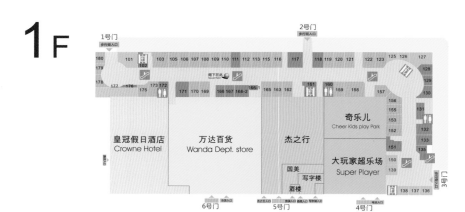

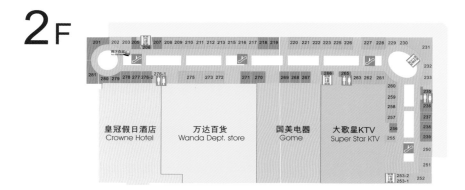

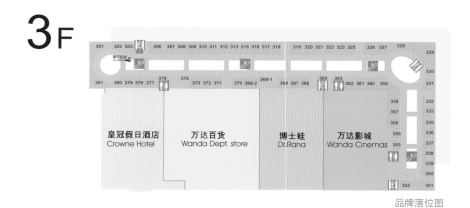

品牌落位图

夜景

宁波江北
万达广场
NINGBO JIANGBEI WANDA PLAZA

总平面图

宁波江北万达广场位于宁波市江北区江北大道以东、云飞路以北，北侧、东侧临宝庆路。商业总建筑面积15.85万平方米，其中地上部分5层，8.75万平方米，地下部分2层，7.1万平方米。业态包括室外商业步行街、室内商业步行街、娱乐楼、国美电器、万达百货及乐购超市。室外商业步行街2层，室内商业步行街3层，娱乐楼4层，万达百货5层。室内商业步行街以圆形和椭圆形中庭为主题，结合生活感较强的建筑形式使总平面上有很强的向心性，吸引聚合大商业的人气。宁波江北万达广场形成"两片、两街"的空间结构。两片——分别为西面的大商业区块和东面的商务楼区块；两街——分别指室内和室外两条紧密联系的"凹"字型商业步行街。在整体的形象塑造上，强调地标性和独特性，营造大尺度的城市景观。高层点式商务楼于地块东侧成片布置，与西侧购物中心、室外商业街相对低矮的建筑形态形成反差，强化了建筑的竖向空间尺度，给人以强烈视觉冲击。

室外日景

3F

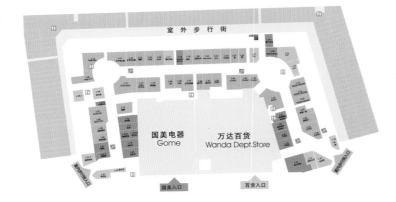

2F

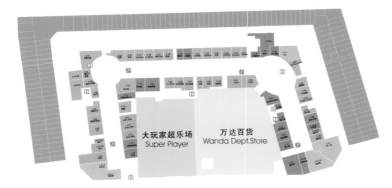

1F

Ningbo Jiangbei Wanda Plaza located in Jiangbei District, Ningbo, East of Jiangbei Avenue, North of Yunfei Road, North and east side next to Baoqing Road. Total commercial floorage is 158,500 m², including 5-storey 87,500 m² above ground, 2-storey 71,000 m² underground. Commercial complex includes an outdoor pedestrian street, indoor pedestrian street, recreational buildings, GOME and Wanda Dept. Store and Tesco supermarkets. Outdoor commercial pedestrian retails are 2-storey high, indoor pedestrian retails are 3-storey high, entertainment building is 4-storey high, and Wanda Dept. Store is 5-storey high. Indoor commercial Street featured with circular and oval-shaped atrium as its theme, combined with the dramatic forms, are highly concentric, attracting great amount flow of the people. Ningbo Jiangbei Wanda Plaza formed "two district, two Street" structure of space. Two districts include the large commercial blocks on the west side and commercial buildings on the East side; Two Streets include Indoor and outdoor two closely linked walking Street. The overall master plan emphasizes landmarks and unique, creating large scale urban landscape. The high tower buildings locate on the east, colse to the shopping center on the west, strengthen the vertical space, give a strong visual impact.

广场夜景

室内步行街内景

室内步行街扶梯

长春红旗街
万达广场
CHANGCHUN HONGQIJIE WANDA PLAZA

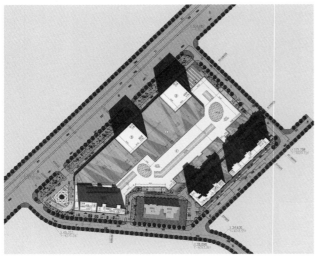

总平面图

长春红旗街万达广场位于长春市四大商圈之一的红旗商圈核心区内，地块东至虎林路，南至同德路，西至红旗街，北至信义路。规划用地面积4.8万平方米，总建筑面积约30万平方米，其中地上建筑面积约26万平方米，地下建筑面积约4万平方米。建筑高度100米，其中地上33层，地下2层。

Changchun Hongqijie Wanda Plaza is located in one of the four big business zones (the Hongqi area). It lies east to the Hulin Road, south to the Tongde Road, west to the Hongqi street, north to the Xinyi Road. The planning area is 48,000m², with a total construction area of about 300,000 m², the ground area of about 260,000 m², underground area of about 40,000 m². Building height is 100 m high; with the ground 33 layers, underground 2 layers.

广场夜景

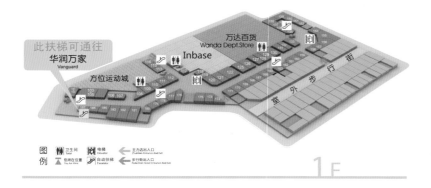

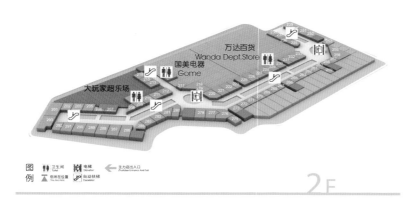

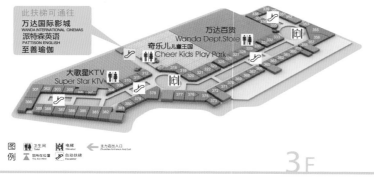

品牌落位图

本项目2010年10月29日盛大开业，其中购物中心由室内步行街、百货楼、娱乐楼、综合楼、大型超市以及与其配套的地下停车场、地下设备用房、管理用房等共同构成，是长春市第一家"城市综合体"。

The project grandly opened on October 29th, 2010. The shopping center consist of indoor pedestrian streets, department stores, entertainment buildings, complex buildings, large supermarkets and underground parking garage, underground equipment rooms and management buildings, which is the first "city complex" in Changchun.

室内步行街

广场入口

沈阳铁西万达广场
SHENYANG TIEXI WANDA PLAZA

沈阳铁西万达广场位于沈阳市铁西区沈辽路北侧，景星南街以东和兴华南街以西之间，沈阳铁西万达广场的总规划用地面积约20万平方米，总建筑面积约91万平方米。其中，东南侧地块为万达商业综合体，总用地面积为5.3万平方米；总建筑面积约29.8万平方米。其中地上部分30层，共23.7万平方米；地下部分2层，6万平方米。建筑高度100米。

Shenyang Tiexi Wanda Plaza located between the east of Jingxing South Street and the west of Xinghua South Street is positioned to the north of Shen-liao Road, Tiexi District, Shenyang. The total site area of this project is about 200,000 m² and total floor area is 910,000 m². The commercial complex is planned in south east plot with 53,000 m² site area and gross floor area of 298,000 m², of which 30 floors above ground with building height of 100 m and area of 237,000 m², and 2 underground floors with area of 60,000 m².

总平面图

广场夜景

购物中心业态品类丰富，主力店包含万达百货、万达影城、连锁家电、大歌星KTV、连锁超市、大玩家超乐场等并结合娱乐休闲类、生活配套类商铺及品牌。

Wanda Tiexi Plaza opens as integrated building complex accommodating shopping center, Wanda Dept. Store, Wanda Cinemas, GOME, Super Star KTV, supermarket, Super Player Park and entertainment facilities, and shopping and service street.

沈阳铁西万达广场形成沈阳新的城市副中心。

Shenyang Tiexi Wanda Plaza has developed as a new sub center of Shenyang.

室内步行街扶梯

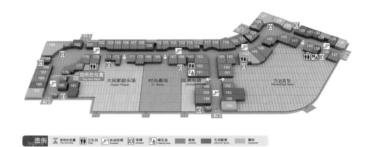

1F 潮流殿堂
Fashion

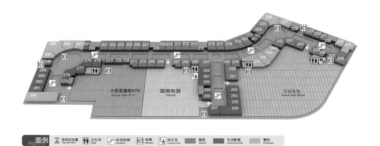

2F 动感磁场
Dynamic

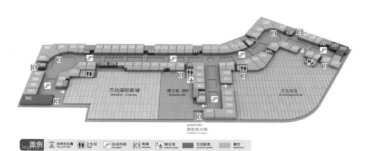

3F 食尚天地
Delicacies

品牌落位图

室内步行街入口

室内步行街

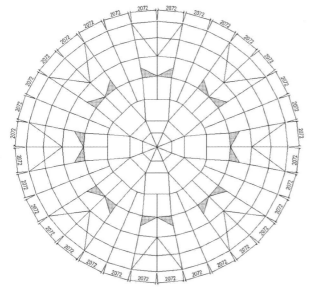

采光顶平面图

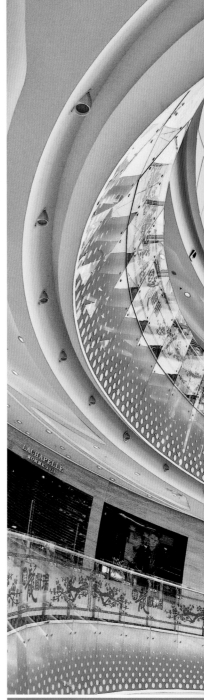

室内步行街中庭

室内步行街扶梯

室内步行街

室外入口夜景

PART 2 酒店
HOTELS

万达集团目前已开业28家五星和超五星级酒店。万达集团计划到2015年开业80家五星和超五星级酒店,营业面积300万平方米,成为全球最大的五星级酒店业主。万达拥有五星级酒店品牌——万达嘉华、超五星级酒店品牌——万达文华以及顶级奢华酒店品牌——万达瑞华。

Wanda Group has opened 28 5-star and super 5-star hotels so far. The Group also plans to open 80 5-star and super 5-star hotels by 2015, with an operation area of 3 million square meters, as the world's largest 5-star hotel owner. Wanda Group owns its 5-star hotel brand — Wanda Realm, super 5-star hotel brand — Wanda Vista and top luxury hotel brand — Wanda Reign.

万达集团旗下酒店分为城市酒店和度假酒店两大类。万达城市类酒店通常位于城市中心,一般和万达广场等业态组成万达城市综合体。万达度假类酒店通常位于文化旅游区,一般和旅游小镇、秀场、主题公园等组成万达文化旅游项目。

Wanda Group's hotels are divided into two major categories: the urban hotels and the resort hotels. The urban hotels are usually located in the city center, normally forming a Wanda urban complex together with Wanda Plaza, etc. The resort hotels are usually located in the cultural tourism zone, normally forming a cultural tourism project together with the tourist town, the show theatre, the theme park and so on.

合肥万达威斯汀酒店
THE WESTIN HEFEI BAOHE

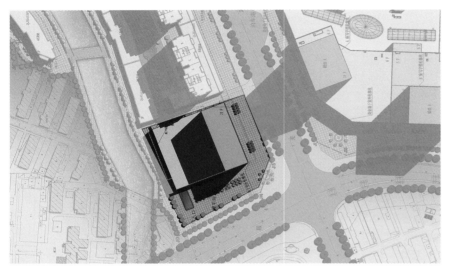

总平面图

合肥威斯汀大酒店位于合肥包河万达广场，万达广场建筑面积约70万平方米，占地面积11.07公顷，位于安徽省合肥市包河区美菱旧厂区地块，东至巢湖路，南芜湖路，西至马鞍山路，北至现状地界。

The Westin Hefei Baohe is located at Baohe Wanda Plaza in Hefei. Wanda Plaza occupies 11.07 ha. and includes a total floor area of 700,000 m². The site is an old plant of Meiling, with Chaohu Road to east, Wuhu Road to south, Maanshan Road to west and existing boundary to north.

酒店总建筑面积约4.82万平方米，其中地上建筑面积3.82万平方米，地下建筑面积1万平方米，于2010年12月23日开业。

The gross floor area of this hotel is 48,200 m², including 38,200 m² above ground area and 10,000 m² basement. The hotel was opened on December 23, 2010.

酒店晚霞

酒店日景

酒店拥有完善的餐饮设施：拥有三家餐厅，包括一家中餐厅，一家特色餐厅，一家全日餐厅。酒店的会议设施位于三层：包括一个无柱大宴会厅（1400平方米）及两个会见厅，12个会议室。大宴会厅可容纳1000人，可分隔成三个部分使用。酒店的康体设施位于四层：包括恒温泳池、健身中心和美发沙龙。

The hotel boasts complete dining facilities, including three restaurants (one Chinese restaurant, one feature restaurant and one all day dining). The meeting facilities are arranged on level 3, including 1 column free big banquet hall (1,400 m² or 1000 seats, can be flexibly divided into 3 parts), 2 meeting halls, and 12 conference rooms. Sports facilities are located on floor 4 which accommodates constant temperature swimming pool, fitness center, and hair salon.

酒店拥有客房标准间316间，普通套房21套，部长套房3套，总统套房1套，此外还设有行政酒廊。

The hotel has 316 standard rooms, 21 common suites, 3 minister suites, 1 president suites as well as executive lounge.

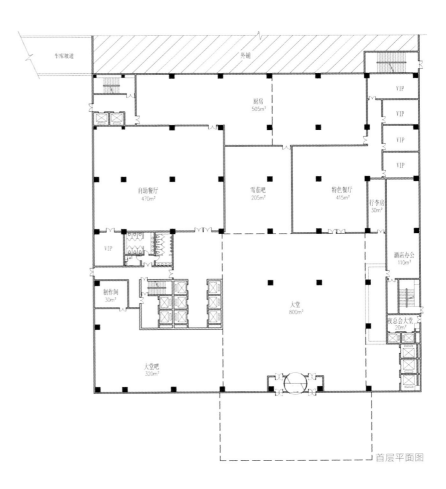

首层平面图

行政层平面图

酒店大堂

襄阳万达
皇冠假日酒店
CROWNE PLAZA XIANGYANG

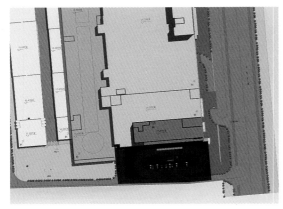

总平面图

襄阳万达皇冠假日酒店于2010年11月26日隆重开业。酒店占地建筑高度86.4米，地上建筑面积34441平方米。楼高21层，其中地上19层，地下2层。投资业主方为大连万达集团股份有限公司襄阳万达广场投资有限公司。聘请目前全球最大及网络分布最广的专业酒店管理集团——洲际酒店管理集团对酒店进行经营管理。

Crown Plaza Xiangyang grandly opened on November 26th, 2010. The hotel building height is 86.4 m, and its building area is 34,441 m^2 with 21 stories, among them, 19-story above ground, 2-story underground. The investment owner is Xiangyang Wanda Plaza investment Co., Ltd. of the Dalian Wanda Group Company who employ the world's largest and most widely distributed network of professional hotel management group — Intercontinental Hotel Management Group to the Hotel's daily business management.

酒店占地面积7211平方米；建筑面积41441平方米（含地下车库7000平方米，容积率6.44）。酒店为1栋19层的建筑，包括客房303间。

The hotel covers an area of 7,211 m^2; construction area of 41,441 m^2 (including underground garage 7,000 m^2, the floor area rate of 6.44). The hotel consists of 1 building, which has 19 stories and 303 guest rooms.

酒店主入口

酒店大堂

大堂休息区

酒店前台

福州万达威斯汀酒店
THE WESTIN FUZHOU MINJIANG

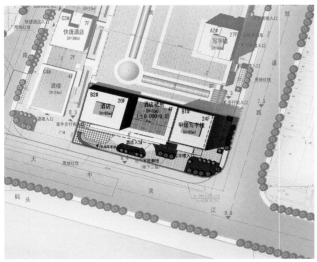

总平面图

福州万达威斯汀酒店坐落于闽江江畔，位于福州金融街万达广场内，毗邻福州市海峡会展中心和著名的鼓山旅游景区，离市中心仅需15分钟车程，交通十分便捷。

The Westin Fuzhou Minjiang is located in Fuzhou Financial Street Wanda Plaza along the Min River, adjacent to Fuzhou Strait Exhibition Center and the famous Gulangyu scenic spot. Transportation is very convenient with only 15 minutes drive from hotel to Fuzhou downtown.

酒店临江平台

酒店主入口

宴会厅

酒店大堂

福州万达威斯汀酒店总建筑面积约4.9万平方米，拥有300多间不同房型的客房，配备天梦之床、天梦之浴等人性化设计；1400余平方米的大宴会厅以及多间中小会议室可以承接各种类型的会议和宴会，配以威斯汀个性化的服务、量身定制的菜单、最新的现代科技，成为商务会议、婚礼宴请与庆典活动的首选场所；位于酒店四层的中餐厅，每间贵宾包厢都配有观景平台，让客人尊享美食的同时领略闽江的无限风光。

The gross floor area of The Westin Fuzhou Minjiang is about 49,000 m², with more than 300 different guest rooms, equipped with the humanistic design of TianMeng Bed and TianMeng Bath; more than 1,400 m² large banquet hall as well as many small and medium sized conference rooms can undertake various meetings and banquets, matching with Westin personalized service, specific tailored menus, and the latest modern technology, make the hotel become the first choice for the business conferences, wedding banquets and celebration activities; The Chinese restaurant located at the fourth floor with VIP room, equipped with viewing platform, can satisfy guests to enjoy both delicious food and the infinite scene of Min River at the same time.

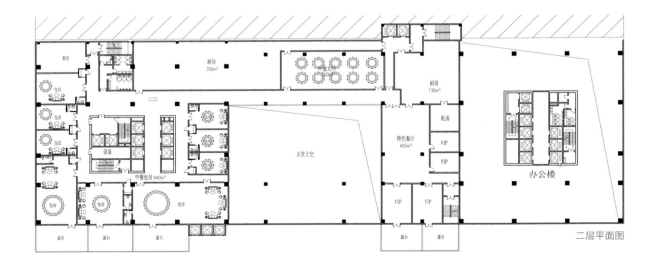

二层平面图

酒店外观日景

室内游泳池

无锡万达喜来登酒店
SHERATON WUXI BINHU HOTEL

无锡万达喜来登酒店坐落于滨湖河埒区的中心地带，毗邻无锡万达广场，占地约3.6公顷，总建筑面积约4.7万平方米，于2010年8月27日正式开业。

Sheraton Wuxi Binhu hotel is located in Helie center zone of Binhu Distrcit, adjacent to Wuxi Wanda Plaza. The hotel covers the site area of about 3.6 ha, with the gross floor area of about 47,000 m². It was opened grandly on August 27th, 2010.

酒店拥有完善的配套设施：酒店一层有西餐厅、红酒吧、全日餐厅。二层设有大型中餐厅，酒店的会议设施位于三层：多功能会议区配有无线上网设施，一流的灯光，音响设备和技术服务，完全能满足各种会议和宴会的需求，包括一个无柱大宴会厅（1500平方米）及6个多功能厅，大宴会厅可容纳1000人，可分隔成三个部分使用。酒店的康体设施位于四层：包括恒温泳池、健身中心和美发沙龙。

The hotel possesses perfect supporting facilities: western restaurant, wine bar and diurnal restaurant on the first floor, a large Chinese restaurant on the second floor, hotel conference facilities on the third floor with the multifunction meeting area equipped with wireless Internet facilities, first-class lighting, audio equipments and technical services, can absolutely meet the needs of all kinds of meetings and banquets, including a large banquet hall without any column (1500 m²) and 6 multifunction halls. The large banquet hall can accommodate 1000 persons, and can be divided into three parts to use. The hotel Health & fitness facilities are located on the fourth floor including constant temperature swimming pool, fitness center and hair salon.

酒店拥有客房标准间342间，普通套房24套，部长套房5套，总统套房1套，此外还设有行政酒廊。

The hotel has 342 standard guest rooms, 24 ordinary suites, 5 minister suites, 1 presidential suite, and also an executive lounge besides that.

酒店广场夜景

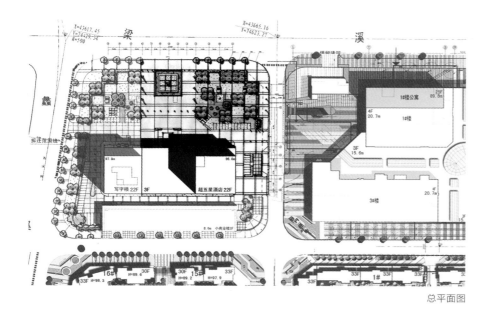

总平面图

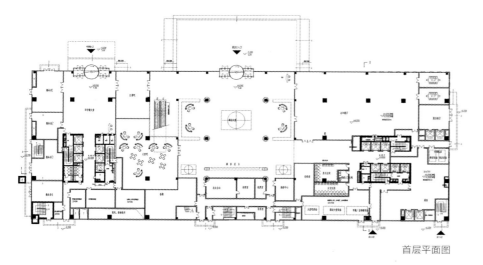

首层平面图

自助餐厅

大堂吧

宜昌万达
皇冠假日酒店
CROWNE PLAZA YICHANG

全景鸟瞰

酒店外立面日景

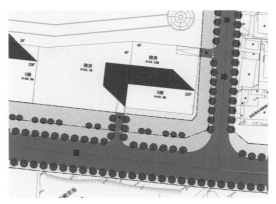

总平面图

酒店入口

宜昌万达皇冠假日酒店坐落于宜昌市伍家岗区，位于沿江大道与夷陵大道之间，项目东临万寿路，南依沿江大道，西靠胜利一路，北至夷陵大道，堪称伍家岗区和西陵区交接的门户位置。总建筑面积3.93万平方米，其中地上建筑面积3.23万平方米。

Crowne Plaza Yichang is located in Wujiagang District in the city of Yichang, situated between Yanjiang Road and Yiling Road. It is closed to Wanshou Road to the east, Yanjiang road to the south, Shengliyi Road to the west, and north to Yiling Road. It is so called as a "Gateway" connecting Wujiagang and Xiling District. Its overall floorage is 39,300 m^2, including the above ground area of 32,300 m^2.

总客房数306间，设有1000平方米大宴会厅和800平方米大堂。

There are 306 guest rooms, a 1,000 m^2 banquet hall, and a 800 m^2 main lobby.

酒店整体外立面以雄伟的姿态与创新的设计内涵集中体现了都市化、现代化的设计理念，内装设计充分挖掘地方文化元素，典雅、大气。景观设计营造静谧、高贵的氛围，并实现了与商业广场的有效分隔与联系。

The whole exterior is designed in a magnificent and innovative way, which represents metropolitan and modernized design concept. Meanwhile, the interior design fully explores local culture elements, which is elegant and geneous. The landscape design creates a sense of peace and nobility. At the same time, it makes effective division and connection with business plaza.

大堂吧

酒店大堂

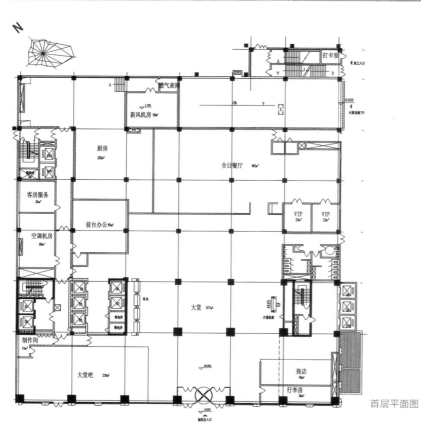
首层平面图

酒店大堂

PART 3 导向标识精选
SELECTION OF GUIDING SIGNS

导向标识系统是引导人们在公共场所活动的综合性公共信息系统。它的基本功能是指引方向，重要辅助功能是强化区域形象。

Way-finding & signage system is the comprehensive public information system that is used to guide activities of people at public places. Its basic function is to guide the direction, with an important auxiliary function of enhancing the image of the region.

万达广场的导向标识系统经过多年的项目实践及总结，已经形成自身较为完善的一套规划设计标准，其中主要包括户外、室内和地下停车场三大部分。

After numerous years of project practice and summarization, the way-finding & signage system of Wanda Plaza has become a complete set of planning and designing norms, mainly including three parts: the outdoor, the indoor and the underground parking space.

设计中标识布点须遵循车、客流行进流线，并结合现场环境设置，保证标识醒目、易识别；访客视觉性通透。

In designing, the signage layout must follow the routes of vehicles flow and passengers flow, and, based on the environmental conditions, guarantee that the signage is eye-catching and easy to recognize, providing visual permeability to visitors.

不同牌体的造型、设计元素、牌体本身的尺度及材质需与广场对应区域的整体设计风格相协调一致并满足各功能性信息的排布要求。

The shape, designing elements, dimension and material quality of different signs shall conform to the overall designing style of the corresponding areas of the Plaza and satisfy the requirement for the configuration of all kinds of functional information.

PART 4 项目索引
INDEX OF THE PROJECTS

项目索引 2010
INDEX OF THE PROJECTS 2010

万达广场/ WANDA PLAZAS

合肥包河万达广场
HEFEI BAOHE WANDA PLAZA
2010.12

福州金融街万达广场
FUZHOU FINANCIAL STREET WANDA PLAZA
2010.12

武汉菱角湖万达广场
WUHAN LINGJIAOHU WANDA PLAZA
2010.12

绍兴柯桥万达广场
SHAOXING KEQIAO WANDA PLAZA
2010.12

广州白云万达广场
GUANGZHOU BAIYUN WANDA PLAZA
2010.12

宁波江北万达广场
NINGBO JIANGBEI WANDA PLAZA
2010.12

宜昌万达广场
YICHANG WANDA PLAZA
2010.11

襄阳万达广场
XIANGYANG WANDA PLAZA
2010.11

天津河东万达广场
TIANJIN HEDONG WANDA PLAZA
2010.11

济南魏家庄万达广场
JINAN WEIJIAZHUANG WANDA PLAZA
2010.11

呼和浩特万达广场
HOHHOT WANDA PLAZA
2010.11

包头青山万达广场
BAOTOU QINGSHAN WANDA PLAZA
2010.11

长春红旗街万达广场
CHANGCHUN HONGQIJIE WANDA PLAZA
2010.10

无锡滨湖万达广场
WUXI BINHU WANDA PLAZA
2010.09

沈阳铁西万达广场
SHENYANG TIEXI WANDA PLAZA
2010.08

酒店/ HOTELS

合肥万达威斯汀酒店
THE WESTIN HEFEI BAOHE
2010.12

福州万达威斯汀酒店
THE WESTIN FUZHOU MINJIANG
2010.12

宜昌万达皇冠假日酒店
CROWNE PLAZA YICHANG
2010.11

襄阳万达皇冠假日酒店
CROWNE PLAZA XIANGYANG
2010.11

无锡万达喜来登酒店
SHERATON WUXI BINHU HOTEL
2010.09

2010
万达商业规划研究院
WANDA COMMERCIAL PLANNING & RESEARCH ISTITUTE CO., L

刘江　尹富庚　李晓燕　王朕　吴迪
王鑫　夏洪兴　边宇　屈娜　王弘成
吴绿野　赵辉　董丽梅　熊伟

谭喆　潘晓光　郭峤宇　吴刚
沈剑锋　莫鑫　王光宇　王群华
谷全　唱亮　江勇　赵沛
张文和　毛晓虎　杨洪海　朱莹洁　任小

李成斌　兰峻文　刘悦飞　曹尧
蓝毅　曾明　张振宇　严铁钰
张涛　赵新宇　曹亚星　郭薇
王明妍　邹成江　杨旭　吴昊

田迎斌　王绍合　王元　冯腾飞　国文　吕永军
文善平　莫力生　刘玉峰　范珑　李峻　孙培宇
袁志浩　徐涛　李峥　孙多斌　黄勇　陈忆秋

唐海江　李杨　孙楠　赖建燕　黄大卫　朱其玮　叶宇峰
雷磊　刘婷　代红　李斌　刘平　谷建芳　王惟　张宝鹏

刘晓　臧久龙　田杰　魏成刚　叶甲刚
马红　刘婷　梁爽　张琳　杨彬　曹冰
于瑞勇　屠波　龙向东　尚海燕

杨世杰　张振宇

彬　童球　孙明杰

周藐　邵汀潇　张鹤　张彬　李楠　吕彬锋　李甜
孙佳宁　曹莹　苗凯峰　于光炤　霍小虎　杨根朝
刘佳　郭双　张华　王燕　李晶　张帅克

余小莉　李文娟　秦好刚　周力大　郝宁克　张飚　万志斌
李兵　阎红伟　李树靖　侯卫华　王魏巍　曾少卿　刘子瑜
温亚玲　胡伟　王巍　梅咏　张帆

酒店主入口雨棚

合肥包河万达广场

图书在版编目（CIP）数据

万达商业规划 2010 / 万达商业规划研究院主编.
—北京：中国建筑工业出版社，2013.5
ISBN 978-7-112-15409-8

Ⅰ.①万… Ⅱ.①万… Ⅲ.①商业区—城市规划—中国
Ⅳ.①TU984.13

中国版本图书馆CIP数据核字(2013)第086894号

责任编辑：徐晓飞　张　明　徐　冉　李　鸽
美术编辑：康　宇
装帧设计：洲联集团·五合国际·五合视觉
责任校对：姜小莲　王雪竹

万达商业规划2010
万达商业规划研究院　主编
*
中国建筑工业出版社出版、发行（北京西郊百万庄）
各地新华书店、建筑书店经销
北京雅昌彩色印刷有限公司制版
北京雅昌彩色印刷有限公司印刷
*
开本：787×1092毫米　1/8　印张：23 $\frac{1}{2}$　字数：643千字
2013年6月第一版　2013年6月第一次印刷
定价：298.00元
ISBN 978-7-112-15409-8
（23408）
版权所有　翻印必究
如有印装质量问题，可寄本社退换
（邮政编码 100037）